PRACTICAL BIOCHEMISTRY

PRACTICAL BIOCHEMISTRY

Second Edition

Geetha Damodaran K
MBBS MD (Biochemistry)

Additional Professor
Department of Biochemistry
Government Tirumala Devaswom (TD) Medical College
Alappuzha, Kerala, India

JAYPEE The Health Sciences Publisher

New Delhi I London I Philadelphia I Panama

 **Jaypee Brothers Medical Publishers (P) Ltd.**

Headquarters
Jaypee Brothers Medical Publishers (P) Ltd.
4838/24, Ansari Road, Daryaganj
New Delhi 110 002, India
Phone: +91-11-43574357
Fax: +91-11-43574314
E-mail: jaypee@jaypeebrothers.com

Overseas Offices

J.P. Medical Ltd.
83, Victoria Street, London
SW1H 0HW (UK)
Phone: +44 20 3170 8910
Fax: +44 (0)20 3008 6180
E-mail: info@jpmedpub.com

Jaypee-Highlights Medical Publishers Inc.
City of Knowledge, Bld. 237, Clayton
Panama City, Panama
Phone: +1 507-301-0496
Fax: +1 507-301-0499
E-mail: cservice@jphmedical.com

Jaypee Medical Inc.
The Bourse
111, South Independence Mall East
Suite 835, Philadelphia, PA 19106, USA
Phone: +1 267-519-9789
E-mail: jpmed.us@gmail.com

Jaypee Brothers Medical Publishers (P) Ltd.
17/1-B, Babar Road, Block-B, Shaymali
Mohammadpur, Dhaka-1207, Bangladesh
Mobile: +08801912003485
E-mail: jaypeedhaka@gmail.com

Jaypee Brothers Medical Publishers (P) Ltd.
Bhotahity, Kathmandu, Nepal
Phone: +977-9741283608
E-mail: kathmandu@jaypeebrothers.com

Website: www.jaypeebrothers.com
Website: www.jaypeedigital.com

© 2016, Jaypee Brothers Medical Publishers

The views and opinions expressed in this book are solely those of the original contributor(s)/author(s) and do not necessarily represent those of editor(s) of the book.

All rights reserved. No part of this publication may be reproduced, stored or transmitted in any form or by any means, electronic, mechanical, photocopying, recording or otherwise, without the prior permission in writing of the publishers.

All brand names and product names used in this book are trade names, service marks, trademarks or registered trademarks of their respective owners. The publisher is not associated with any product or vendor mentioned in this book.

Medical knowledge and practice change constantly. This book is designed to provide accurate, authoritative information about the subject matter in question. However, readers are advised to check the most current information available on procedures included and check information from the manufacturer of each product to be administered, to verify the recommended dose, formula, method and duration of administration, adverse effects and contraindications. It is the responsibility of the practitioner to take all appropriate safety precautions. Neither the publisher nor the author(s)/editor(s) assume any liability for any injury and/or damage to persons or property arising from or related to use of material in this book.

This book is sold on the understanding that the publisher is not engaged in providing professional medical services. If such advice or services are required, the services of a competent medical professional should be sought.

Every effort has been made where necessary to contact holders of copyright to obtain permission to reproduce copyright material. If any have been inadvertently overlooked, the publisher will be pleased to make the necessary arrangements at the first opportunity.

Inquiries for bulk sales may be solicited at: jaypee@jaypeebrothers.com

Practical Biochemistry

First Edition: 2011
Second Edition: **2016**
ISBN: 978-93-5152-994-1
Printed at: Sanat Printers

Dedicated to

My father Sri KV Damodaran

Preface to the Second Edition

The motive behind writing *Practical Biochemistry* was to make the practical concepts of biochemistry clear for medical students in an examination-oriented format. I am happy to understand that it has almost fulfilled the purpose.

It is time now to launch the second edition of the book. The book is primarily written for medical students, but it should also be useful to students of dentistry and pharmacy.

In this edition, most of the diagrammatic figures are replaced by real images in Qualitative Analysis and Spotters sections. This will definitely impart a realistic feeling of learning qualitative tests. More questions are added to Spotters section.

Organization of the Book: It is done in an examination-oriented manner. Usually, students have to go through four or five modes of practical examinations with specific intentions. Hence, the contents of the book are organized accordingly, to facilitate the learning process and preparation for the practical examination.

Section on Qualitative Analysis: To provide comprehensive analysis of different tests in order to impart a grasp of concepts to learners.

Section on Quantitative Analysis: To enable the students to master estimation experiments. The various aspects such as different methods, reference ranges and interpretations of each item will help to achieve this.

Section on Charts: To improve the critical thinking, an essential quality for any doctor or scientist. This section contains 54 clinically-oriented questions with answers to reinforce the information tested by the questions.

Section on Spotters: To sharpen the abruptness, a skill to be attained by a medical professional. More than 100 spotters are included to serve this purpose.

Section on Objective Structured Practical Examination (OSPE) Questions: To improve the clarity of practical work, this section also contains model OSPE questions for practice. This section will definitely boost the confidence of students to face OSPE.

The chapters for experiments contain, sub-section *Questions* to improve comprehension and also sub-section *Reagent Preparation*, which is useful for students to understand the ingredients of various reagents and for the laboratory staff to prepare reagents required for different experiments.

I warmly welcome all suggestions from students and faculties for improving the subsequent editions of the book. Please send your suggestions at *biochempra@gmail.com*.

Geetha Damodaran K

Preface to the First Edition

Nearly two decades of teaching experience have driven me to write the book. During these years, I realized that if an illustrated book is available, then it will be easy for students to recollect the experiments done earlier and to face the different types of questions during practical examinations. Hence, all the items in the book are illustrated.

The contents of the book are structured in the practical examination-oriented manner. The major sections are qualitative experiments, quantitative experiments, charts, spotters and objective structured practical examination (OSPE) questions. All the tests are provided with diagrams and interpretations. This will help the students to understand each concept thoroughly and enable them to use it as an instant doubt-clearing book. I hope that it will be very useful for day-to-day studies and examination preparations.

Details of reagent preparations given along with the respective chapters are useful for the staff involved in the laboratory preparation of practical sessions. This part will also help to improve the level of understanding of students about the reagents, they are using for various experiments in the laboratory.

Questions provided with the chapters are useful for having better clarity and grasp of the topic. Moreover, it will definitely boost the confidence of students to face the examination. Chapters on charts and spotting and OSPE questions are useful for self-training of such type of evaluation methods.

I warmly welcome the views of those using the book and I shall be grateful to the readers for bringing to my notice of mistakes for corrections, in future edition of the book.

Geetha Damodaran K

Acknowledgments

First of all, I thank the supreme power, God for enabling me to do this work. I am always indebted to my parents and teachers for molding me to reach at least this level. I extend my gratitude to my colleagues for their support. I should thank my husband Dr PK Balachandran for constantly persuading me to write. Moreover, my special thanks to the team of M/s Jaypee Brothers Medical Publishers (P) Ltd, New Delhi, India, for their expertise in publication.

Contents

SECTION 1: QUALITATIVE ANALYSIS

1. Reactions of Carbohydrates — 3
Classification   3
Reactions of Monosaccharides   4
Reactions of Disaccharides   11
Reactions of Polysaccharides   15
Reactions of Starch   15
Identification of Unknown Carbohydrates   17
Questions   18
Reagent Preparation   18

2. Reactions of Proteins — 19
Precipitation Reactions of Proteins   19
Color Reactions of Proteins and Amino Acids   25
Specific Color Reactions used to Identify the Side Chain (R) Groups of Amino Acids   27
Reactions of Albumin   32
Reactions of Casein   32
Identification of Unknown Proteins   34
Reactions of Gelatin   34
Questions   35
Reagent Preparation   35

3. Reactions of Lipids — 37
General Reactions of Lipids   37
Reactions of Fats and Fatty Acids   37
Cholesterol   39
Questions   41
Reagent Preparation   41

4. Reactions of Urea — 42
General Reactions of Urea   42
Questions   44
Reagent Preparation   44

5. Reactions of Creatinine — 45
General Reactions of Creatinine   45
Questions   46
Reagent Preparation   46

6. Reactions of Uric Acid — 47
General Reactions of Uric Acid 47
Questions 49
Reagent Preparation 49

7. Schemes for Identification of Biologically Important Compounds — 50
For Biologically Important Carbohydrate Substances Comprising Glucose, Fructose, Sucrose, Lactose, Maltose 50
For Biologically Important Substances Comprising Glucose, Fructose, Sucrose, Lactose, Maltose, Albumin, Urea, Uric Acid and Creatinine 51
For Biologically Important Substances Comprising Glucose, Fructose, Sucrose, Lactose, Maltose, Albumin, Casein, Urea, Uric Acid and Creatinine 52

8. Urine Analysis — 53
Analysis of Normal Constituents of Urine 53
Analysis of Abnormal Constituents of Urine 60
Questions 66
Reagent Preparation 66

9. Identification of Hemoglobin Derivatives — 67
Spectroscopic Examination of Hemoglobin Pigments 67
Identification of Hemoglobin Pigment by Microscopy 71
Questions 72
Reagent Preparation 72

10. Reactions of Milk — 74
Precipitation of Casein from Milk 74
Questions 75
Reagent Preparation 75

SECTION 2: QUANTITATIVE ANALYSIS

11. Introduction to Quantitative Analysis — 79
Principles of Colorimetry 79
Components of Photoelectric Colorimeter 83
Measurement in a Photoelectric Colorimeter 83
Questions 84

12. Determination of Glucose — 85
Determination of Glucose Concentration 85
Glucose Tolerance Test 88
Questions 91
Reagent Preparation 91

13. Determination of Urea — 93
Urea 93
Questions 96
Reagent Preparation 96

14. Determination of Creatinine — 98
Creatinine 98
Questions 101
Reagent Preparation 101

15. Determination of Total Protein and Albumin — 103
Methods of Protein Estimation 103
Determination of Albumin 104
Questions 105
Reagent Preparation 106

16. Determination of Total Cholesterol — 107
Cholesterol 107
Questions 109
Reagent Preparation 109

17. Determination of Uric Acid — 110
Uric Acid 110
Questions 112
Reagent Preparation 112

18. Determination of Bilirubin — 114
Methods 114
Interpretation 116
Jaundice 117
Questions 117
Reagent Preparation 117

19. Determination of Transaminases — 118
Transaminases 118
Questions 122
Reagent Preparation 122

20. Determination of Alkaline Phosphatase — 124
Methods 124
Isoenzymes of Alkaline Phosphatase 126
Questions 126
Reagent Preparation 126

21. Determination of Total Calcium — 128
Calcium 128
Questions 131
Reagent Preparation 132

22. Determination of Phosphate — 133
Questions 136
Reagent Preparation 136

SECTION 3: CHARTS

23. Charts   141
- Acid Base Disorders *141*
- Jaundice *145*
- Diabetes Mellitus *149*
- Inborn Errors of Metabolism *155*
- Porphyrias *161*
- Vitamins and Minerals *163*
- Tumor Markers *166*
- Water and Electrolytes *167*
- Kidney Diseases *168*
- Biochemical Diagnosis of Myocardial Infarction *170*
- Enzymology *172*

SECTION 4: SPOTTERS

24. Spotters   177
- Instruments *177*
- Reagents *183*
- Indicators *185*
- Crystals *186*
- Separation Techniques *188*
- Graphs *191*
- Tests *195*
- Nutrition *203*
- Spectroscopy *207*
- Conceptual Questions *208*

SECTION 5: OBJECTIVE STRUCTURED PRACTICAL EXAMINATION

25. Objective Structured Practical Examination: Model Questions   221
- General Guidelines *221*

Index   *241*

SECTION 1

Qualitative Analysis

CHAPTER 1

Reactions of Carbohydrates

INTRODUCTION

Carbohydrates are aldehyde or ketone derivatives of polyhydric alcohols. They are widely distributed in plants and animals. Plants synthesize glucose by photosynthesis and it is converted mainly to storage form, the starch and structural framework form, the cellulose.

Animals largely depend on plant sources to obtain carbohydrates though they can synthesize carbohydrates from non-carbohydrate sources like lactate, glycerol and glucogenic amino acids in their body by a pathway called gluconeogenesis.

The glucose is the major form of carbohydrate absorbed from the gut in humans.

According to the metabolic status of the body, glucose has different fates:
- Catabolized to release energy
- Polymerized to form the storage fuel, the glycogen
- Sometimes converted to other sugars like fructose and galactose.

Different types of carbohydrates are present in intracellular and extracellular fluids and are excreted in urine when the concentrations of them rise in the blood in certain diseases, e.g.:

- Diabetes mellitus → glucose in urine
- Fructosuria → fructose in urine
- Galactosemia → galactose in urine.

Hence, it is essential to understand the tests for their detection.

The classification **of carbohydrates** will be useful for the detection of various types of carbohydrates by different chemical tests.

CLASSIFICATION

- **Monosaccharides:** Cannot be hydrolyzed into simpler carbohydrates. They are classified into trioses, tetroses, pentoses, hexoses, heptoses based on the number of carbon atoms present in them. They are again divided into aldoses and ketoses based on the functional group present in them (Table 1.1).
- **Disaccharides:** Give rise to two monosaccharide units upon hydrolysis, e.g.:
  - Sucrose (glucose + fructose)
  - Lactose (glucose + galactose)
  - Maltose (glucose + glucose).
- **Oligosaccharides:** Yield less than ten monosaccharides upon hydrolysis, e.g.:
  - Maltotriose (3 glucose units)
  - Raffinose (glucose + fructose + galactose).

TABLE 1.1 Classification of monosaccharides		
Monosaccharides	Aldoses	Ketoses
Trioses ($C_3H_6O_3$)	Glycerose	Dihydroxyacetone
Tetroses ($C_4H_8O_4$)	Erythrose	Erythrulose
Pentoses ($C_5H_{10}O_5$)	Ribose	Ribulose
Hexoses ($C_6H_{12}O_6$)	Glucose	Fructose

- **Polysaccharides:** Contain more than ten monosaccharide units
  - *Homopolysaccharides* (consisting of same type of monomeric units), e.g.:
    - Polymer of glucose: Starch, glycogen, cellulose
    - Polymer of fructose: Inulin.
  - *Heteropolysaccharides* (consisting of different types of monomeric units).

Glycosaminoglycans (Proteoglycans) e.g.:
- Heparin (sulfated glucosamine + sulfated iduronic acid)
- Hyaluronic acid (β glucuronic acid + N-acetylglucosamine).

REACTIONS OF MONOSACCHARIDES

Monosaccharides possess one or more hydroxyl groups and an aldehyde or keto group. Therefore, many reactions of monosaccharides are the known reactions of alcohols, aldehydes or ketones. Many of the reactions shown by monosaccharides are exhibited by higher carbohydrates also. Differences in the structures of sugars often affect the rate of a reaction and sometimes the ability to react.

The reactions described below, are applied in the identification of sugars.

The reactions due to hydroxyl group: Dehydration (e.g. Molisch test, Rapid furfural test, Seliwanoff's test).

The reactions due to carbonyl group: Reduction (e.g. Benedict's test, Barfoed's test).

Molisch Test (α-naphthol Reaction) (Fig. 1.1)

Procedure: To 3 mL of sugar solution in a test tube, add two drops of Molisch reagent. Mix thoroughly. Add 3 mL of concentrated sulfuric acid along the sides of the test tube by slightly inclining the tube, thus forming a layer of acid (acid being heavier goes down beneath the sugar solution) in the lower part.

Observation: A **violet/purple colored ring** appears at the junction of two liquids.

Principle: Concentrated acid dehydrates the sugar to form furfural (in the case of pentoses) or furfural derivatives (in the case of hexoses and heptoses), which then condense with α-naphthol to give a **reddish violet** colored complex.

Inference: Indicates presence of a carbohydrate and hence the presence of a monosaccharide.

Application of the test: Used as a general test to detect carbohydrate.

Aberrant Observations

- Instead of a violet ring in the Molisch test, appearance of a **dark brown color** indicates **charring of sugar** due to the **heat generated during the addition of acid** (acid water interaction generates heat). It will become obvious when the concentration of the sugar solution is high. To avoid charring, dilute the sugar sample solution with water as illustrated in Figure 1.2 and repeat the Molisch test.
- Appearance of a **green color** while doing the test, which persist even after completion

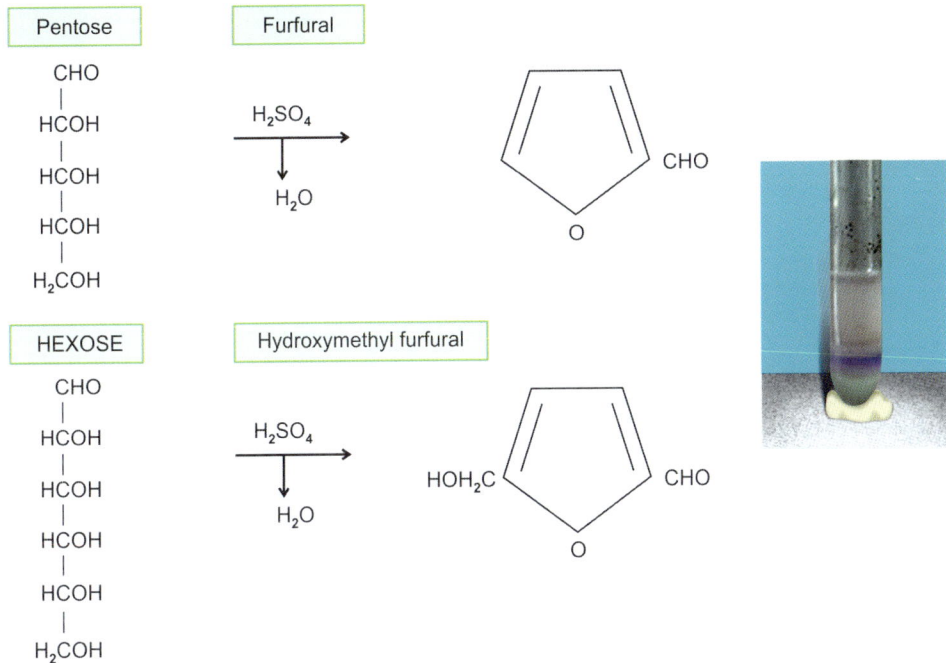

Figure 1.1 Chemistry of Molisch test

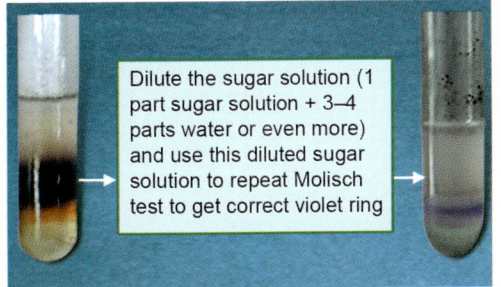

Figure 1.2 Method to avoid charring (Molisch test)

of the test suggest excess use of Molisch reagent than required or due to the presence impurities in the acid.

Benedict's Test (Fig. 1.3)

Procedure: To 5 mL of Benedict's reagent in a test tube add exactly 8 drops of the sugar solution. Mix well. Boil the solution vigorously for two minutes or place in a boiling water bath for three minutes. Allow the contents to cool by keeping in a test tube rack. Do not hasten cooling by immersion in cold water.

Observation: The entire body of the solution will be filled with a precipitate, the color of which varies with the concentration of the sugar solution—green, yellow, orange or brick red.

In the absence of a reducing substance, blue color of the Benedict's reagent remains as such. **The test is sensitive up to 0.1–0.15 g % of sugar in solution (that is Benedict's will not be positive with solutions containing less than 0.1–0.15 g % of sugar).**

Inference: Reducing monosaccharides glucose, fructose, galactose and mannose give a positive reaction with Benedict's reagent.

The color of the precipitate gives an idea about the concentration of the sugar solution as shown in Figure 1.3. Thus, Benedict's test is described as a semi-quantitative test.

| –ve Benedict's test: absence of reducing sugar | +ve Benedict's test: green ~ up to 0.5 g % | +ve Benedict's test: yellow ~ > 0.5–1.0 g % | +ve Benedict's test: orange ~ > 1.0–2.0 g % | +ve Benedict's test: brick red ~ ≥ 2 g % |

Figure 1.3 Benedict's test at different sugar concentrations

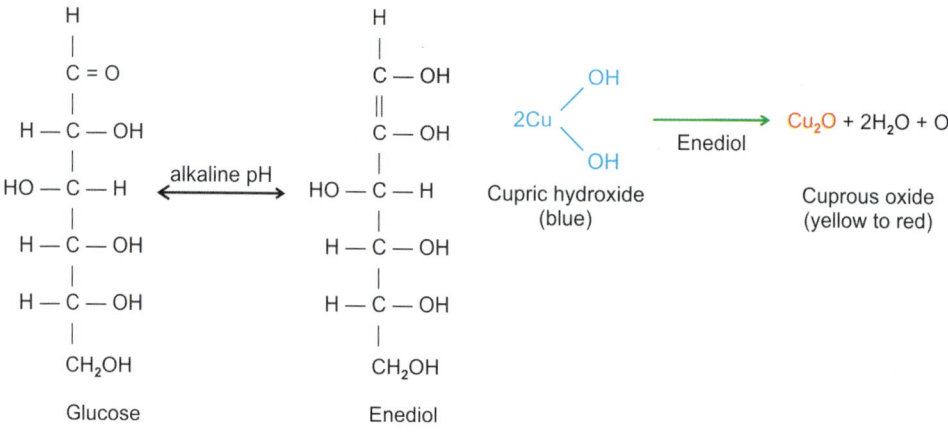

Figure 1.4 Chemistry of Benedict's test

Principle: (Fig. 1.4) Carbohydrates with a free aldehyde or keto group have the ability to reduce various metallic ions. In this test, cupric ions are reduced to cuprous ions by the enediols formed from sugars in the alkaline medium of Benedict's reagent.

Benedict's reagent contains copper sulfate, sodium citrate and sodium carbonate.

Copper sulfate dissociates to give sufficient cupric ions (in the form of cupric hydroxide) for the reduction reactions to occur.

Sodium citrate keeps the cupric hydroxide in solution without getting precipitated (stabilizer).

Sodium carbonate (Na_2CO_3) makes the pH of the medium alkaline.

In the alkaline medium, sugars form enediols, which are powerful reducing agents. They reduce blue cupric hydroxide to insoluble red cuprous oxide.

Application of the test: To detect reducing sugars. It is widely used in detecting glucose in urine even though not specific for glucose.

Barfoed's Test (Fig. 1.5)

Procedure: To 5 mL of Barfoed's reagent in a test tube add 0.5 mL of sugar solution. Mix well. Keep in a boiling water bath for **2 minutes**. Keep the tube in a test tube rack and examine for precipitate.

Observation: A **red** precipitate clinging to the bottom most part of the test tube indicates presence of a monosaccharide.

Inference: The test is answered by monosaccharides only, e.g.: glucose, fructose, galactose, mannose.

Principle: It is a reduction test. Reducing property is due to the carbonyl group (aldehyde or keto group). Barfoed's reagent is copper acetate in acetic acid.

Difference between Barfoed's test and Benedict's test: Barfoed's test differs from Benedict's test with respect to the pH of the medium. It is alkaline in the case of Benedict's where as acidic in the case of Barfoed's. In the acid medium, monosaccharides enolize much more readily than disaccharides and these enediols reduce cupric ions released by copper acetate of Barfoed's reagent.

Points to Ponder

- It is important to keep the time limit (2 minutes) prescribed for Barfoed's test otherwise disaccharides will also respond to the test positively.
- Disaccharides when present in high concentrations (> 5 g%) also will give positive response.
- Unlike the Benedict's test, Barfoed's test is unsuitable for testing sugars in urine or any fluids containing chloride.
- The red precipitate is formed at the bottom of the tube. To see the precipitate, lift the tube to the eye level, otherwise the red precipitate adhering to the bottom most part of the tube may escape notice.

Application of the test: Useful to distinguish between monosaccharides and disaccharides.

Chemistry of the test: Reduction reaction as shown under Benedict's test.

Rapid Furfural Test

Procedure: To 2 mL of concentrated HCl, add 8 drops of sugar solution and 1–2 drops of Molisch reagent. Mix well and heat just to boil.

Observation: Positive reaction is indicated by the development of dark **violet** color (Fig. 1.6).

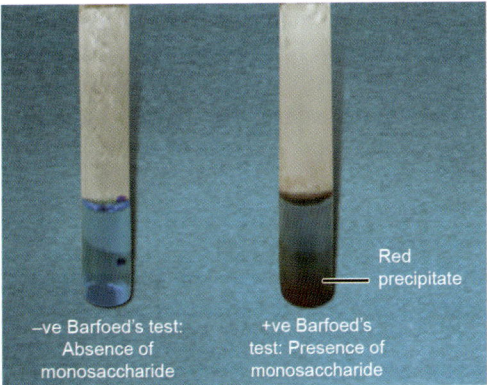

Figure 1.5 Barfoed's test

Figure 1.6 Rapid furfural test

Figure 1.7 Method to avoid charring in rapid furfural test

Inference: Development of violet color within 30 seconds of boiling indicates presence of a keto sugar, e.g. fructose.

Principle: A dehydration reaction due to the hydroxyl groups of the sugar. Concentrated HCl being weaker than concentrated sulfuric acid, dehydrates ketoses (e.g. fructose) more readily than aldoses (e.g. glucose) to form hydroxymethyl furfural, which then condenses with α-naphthol to form a violet colored complex.

Chemistry of the test: Dehydration reaction as shown under Molisch test.

Aberrant reaction: If red color develops instead of violet color due to charring action of acid, dilute the sugar sample with water and conduct the test with diluted sugar solution (Fig. 1.7).

Application of the Test
- For the detection of ketoses
- Useful for differentiating ketoses from aldoses.

Seliwanoff's Test

Procedure: To 3 mL of Seliwanoff's reagent in a test tube add 5 drops of fructose solution and heat the contents to **just boiling**.

Figure 1.8 Seliwanoff's test

Observation: Positive reaction gives a **red color** within 30 seconds (Fig. 1.8).

Inference: This test is given by ketoses, e.g. fructose.

Principle: A dehydration reaction due to the hydroxyl groups of the sugar. Seliwanoff's reagent is resorcinol in dilute hydrochloric acid. Ketoses (e.g. fructose) are more readily dehydrated by HCl than the aldoses to form hydroxymethyl furfural, which then condenses with resorcinol of Seliwanoff's reagent to form a **red colored** complex.

Points to Ponder

- The test is sensitive up to 0.1 g % of fructose in the absence of glucose.
- In the presence of glucose, the test becomes less sensitive to fructose.
- Large amounts of glucose give the same (red) color.
- If the boiling is prolonged, a positive reaction may occur with glucose because of Lobry de Bruyn–van Ekenstein transformation of glucose into fructose, in the presence of acid.

The precautions to be followed to get a positive test for fructose are given below:
- Concentration of HCl used must be less than 12%
- The reaction must be observed **within 20–30 seconds of performing the test**
- Those reactions occurring after 20–30 minutes of boiling must not be considered positive
- Glucose must not be present in amounts more than 2 g% or else it will interfere with the test.

Foulger's Test

Procedure: To 3 mL of Foulger's reagent add 8 drops of sugar solution. Boil for 45 seconds directly on a flame. Allow to cool slowly.

Figure 1.9 Foulger's test

Observation: Deep blue color develops (Fig. 1.9).

Inference: This test is given by ketoses, e.g. fructose.

Principle: A dehydration reaction due to the hydroxyl groups of the sugar. Foulger's reagent contains stannous chloride, urea and 40% H_2SO_4. Ketohexoses form hydroxymethyl furfural with acid which then condenses with stannous chloride and urea to form deep blue color.

Osazone Test

Procedure: To 5 mL of sugar solution in a test tube add 300 mg (one or two scoopfuls) of phenyl hydrazine mixture. Shake well. Heat in a boiling water bath for 15–45 minutes (duration of keeping time in water bath depends on the type of sugar). Then take the tube out of the water bath and allow cooling at room temperature by placing in the test tube rack. Avoid showing under the tap water because rapid cooling disturbs crystallization whereas slow cooling ensures crystallization.

Observation: Crystals are formed readily (within 1–5 minutes) at room temperature in the case of mannose. For other sugars, the minimum incubation time required in minutes in the boiling water bath for the formation of insoluble yellow osazone is given in Table 1.2. Look under the microscope to view the crystals (Fig. 1.10).

TABLE 1.2 Time taken by different monosaccharides for osazone formation	
Monosaccharides	*Time (Minutes)*
Glucose	5
Fructose	2
Galactose	20

Glucose

CHO
|
H — C — OH
|
HO — C — H
|
H — C — OH
|
H — C — OH
|
CH$_2$OH

Fructose

CH$_2$OH
|
C = O
|
HO — C — H
|
H — C — OH
|
H — C — OH
|
CH$_2$OH

Mannose

CHO
|
OH — C — H
|
HO — C — H
|
H — C — OH
|
H — C — OH
|
CH$_2$OH

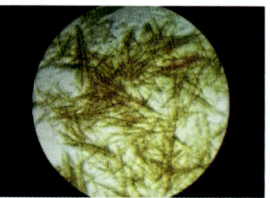

Osazone crystals of glucose, fructose and mannose are needle shaped but when clustered look like a broom or a bundle of hay

Glucose, fructose, mannose yield the same shaped phenyl osazone crystals because of the elimination of differences in configuration about the carbon atoms 1 and 2 during osazone formation.

Figure 1.10 Reason for getting the same shaped osazone crystals for glucose, fructose and mannose

CHO
|
H — C — OH   + H$_2$N.NHC$_6$H$_5$ ⟶   C = N.NHC$_6$H$_5$
| |
HO — C — H   Phenyl hydrazine   H — C — OH   + H$_2$O
| |
R   HO — C — H
Glucose |
                                               R
                                             Glucose phenylhydrazone

C = N.NHC$_6$H$_5$
|
H — C — OH   + 2H$_2$N.NHC$_6$H$_5$ ⟶   C = N.NHC$_6$H$_5$
| |
HO — C — H   phenyl hydrazine   H — C = N.NHC$_6$H$_5$ + C$_6$H$_5$NH$_2$ + NH$_3$
| |
R   HO — C — H   Aniline
Glucose phenylhydrazone |
                                               R
                                             Glucosazone

Figure 1.11 Chemistry of osazone test

Inference: Glucose, fructose, mannose yield the same shaped phenyl osazone crystals because of the elimination of differences in configuration about the carbon atoms 1 and 2 during osazone formation (Fig. 1.11).

Principle: The reaction involves the carbonyl carbon (either aldehyde or ketone as the case may be) and the adjacent carbon. One molecule of sugar reacts with one molecule of phenyl hydrazine initially, to form phenylhydrazone

Points to Ponder

- The test is sensitive up to 0.1 g % of fructose in the absence of glucose.
- In the presence of glucose, the test becomes less sensitive to fructose.
- Large amounts of glucose give the same (red) color.
- If the boiling is prolonged, a positive reaction may occur with glucose because of Lobry de Bruyn–van Ekenstein transformation of glucose into fructose, in the presence of acid.

The precautions to be followed to get a positive test for fructose are given below:
- Concentration of HCl used must be less than 12%
- The reaction must be observed **within 20–30 seconds of performing the test**
- Those reactions occurring after 20–30 minutes of boiling must not be considered positive
- Glucose must not be present in amounts more than 2 g% or else it will interfere with the test.

Foulger's Test

Procedure: To 3 mL of Foulger's reagent add 8 drops of sugar solution. Boil for 45 seconds directly on a flame. Allow to cool slowly.

Figure 1.9 Foulger's test

Observation: Deep blue color develops (Fig. 1.9).

Inference: This test is given by ketoses, e.g. fructose.

Principle: A dehydration reaction due to the hydroxyl groups of the sugar. Foulger's reagent contains stannous chloride, urea and 40% H_2SO_4. Ketohexoses form hydroxymethyl furfural with acid which then condenses with stannous chloride and urea to form deep blue color.

Osazone Test

Procedure: To 5 mL of sugar solution in a test tube add 300 mg (one or two scoopfuls) of phenyl hydrazine mixture. Shake well. Heat in a boiling water bath for 15-45 minutes (duration of keeping time in water bath depends on the type of sugar). Then take the tube out of the water bath and allow cooling at room temperature by placing in the test tube rack. Avoid showing under the tap water because rapid cooling disturbs crystallization whereas slow cooling ensures crystallization.

Observation: Crystals are formed readily (within 1-5 minutes) at room temperature in the case of mannose. For other sugars, the minimum incubation time required in minutes in the boiling water bath for the formation of insoluble yellow osazone is given in Table 1.2. Look under the microscope to view the crystals (Fig. 1.10).

TABLE 1.2 Time taken by different monosaccharides for osazone formation	
Monosaccharides	*Time (Minutes)*
Glucose	5
Fructose	2
Galactose	20

Glucose	Fructose	Mannose
CHO | H—C—OH | HO—C—H | H—C—OH | H—C—OH | CH$_2$OH	CH$_2$OH | C=O | HO—C—H | H—C—OH | H—C—OH | CH$_2$OH	CHO | OH—C—H | HO—C—H | H—C—OH | H—C—OH | CH$_2$OH

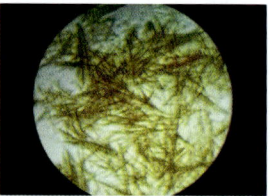

Osazone crystals of glucose, fructose and mannose are needle shaped but when clustered look like a broom or a bundle of hay

Glucose, fructose, mannose yield the same shaped phenyl osazone crystals because of the elimination of differences in configuration about the carbon atoms 1 and 2 during osazone formation.

Figure 1.10 Reason for getting the same shaped osazone crystals for glucose, fructose and mannose

CHO   C=N.NHC$_6$H$_5$
\| \|
H—C—OH   + H$_2$N.NHC$_6$H$_5$ ⟶   H—C—OH   + H$_2$O
\| Phenyl hydrazine   \|
HO—C—H   HO—C—H
\| \|
R   R
Glucose   Glucose phenylhydrazone

C=N.NHC$_6$H$_5$ C=N.NHC$_6$H$_5$
\| \|
H—C—OH   + 2H$_2$N.NHC$_6$H$_5$ ⟶ H—C=N.NHC$_6$H$_5$ + C$_6$H$_5$NH$_2$ + NH$_3$
\| phenyl hydrazine   \|
HO—C—H   HO—C—H   Aniline
\| \|
R   R
Glucose phenylhydrazone   Glucosazone

Figure 1.11 Chemistry of osazone test

Inference: Glucose, fructose, mannose yield the same shaped phenyl osazone crystals because of the elimination of differences in configuration about the carbon atoms 1 and 2 during osazone formation (Fig. 1.11).

Principle: The reaction involves the carbonyl carbon (either aldehyde or ketone as the case may be) and the adjacent carbon. One molecule of sugar reacts with one molecule of phenyl hydrazine initially, to form phenylhydrazone which then reacts with two additional phenyl hydrazine molecules to form the osazones as shown in the Figure 1.11.

Points to Ponder

If the solution appears red after heating process, it indicates that the solution has become concentrated in the boiling process and no crystals will separate in the concentrated form. In such occasion, dilute with water for the separation of crystals.

REACTIONS OF DISACCHARIDES

Introduction

Disaccharides are glycosides in which both components are monosaccharides. The general formula of common disaccharides is $C_{12}H_{22}O_{11}$ The common disaccharides studied are detailed below.

Maltose (α-D-Glucopyranosyl-(1→4) α-D-Glucopyranose) (Fig. 1.12): Maltose yields two glucose molecules upon hydrolysis. Maltose is formed from the hydrolysis of starch by the action of the enzyme maltase. It is also produced as an intermediate product of hydrolysis of starch by mineral acid. It is dextrorotatory, exhibits mutarotation, and reduces metallic ions in alkaline solutions. Like other disaccharides maltose is hydrolyzed by dilute acid leading to the formation of two molecules of glucose. With phenyl hydrazine maltose forms maltosazone.

Examples for other disaccharides that produce only glucose upon hydrolysis:
- *Cellobiose* a β glucoside with 1,4 linkage derived from partial hydrolysis of cellulose.
- *Gentiobiose,* a β glucoside with 1,6 linkage derived from roots of Gentiana lutea
- *Trehalose,* an α glucoside with 1,1 linkage obtained from yeast and mushrooms.

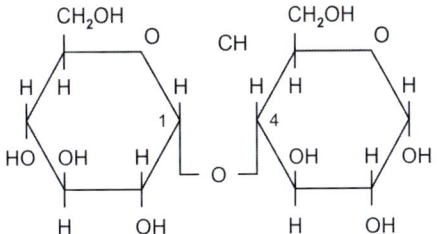

Figure 1.12 Maltose (α-D-Glucopyranosyl-(1→4) α-D-Glucopyranose, i.e. α 1, 4 linkage)

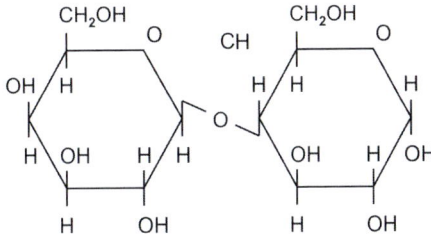

Figure 1.13 Lactose (β-D-Galactopyranosyl-(1→4) β-D-Glucopyranose, i.e. β 1,4 glycosidic linkage)

- *Isomaltose,* an α glucoside with 1,6 linkage formed as a side product of hydrolysis of starch by amylase enzyme.

Lactose (β-D-Galactopyranosyl-(1→4) β-D-Glucopyranose) (Fig. 1.13): Lactose gives rise to one molecule of glucose and one molecule of galactose upon enzymatic (lactase) or acid hydrolysis. Lactose normally present in milk and in the urine of women during latter half of pregnancy and during lactation.

It is dextrorotatory, shows mutarotation in solution. It reduces metallic ions, forms lactosazone with phenyl hydrazine. It is a galactoside since the carbon number 1 of galactose is involved in the β galactosidic bond with the carbon number 4 of glucose.

Sucrose (α-D-Glucopyranosyl-β-D-Fructofuranoside) (Fig. 1.14): Hydrolysis of sucrose yields one molecule of glucose and one molecule of fructose. Sucrose is dextrorotatory. After

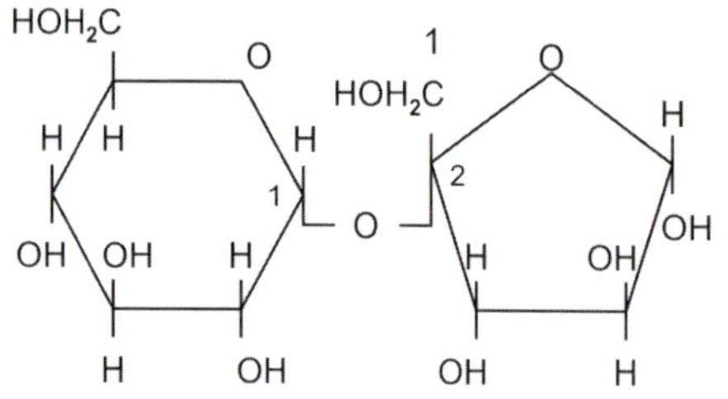

Figure 1.14 Sucrose (α-D- glucopyranosyl (1→2)- β-D-fructofuranoside)

hydrolysis by enzymes or weak acids, it becomes levorotatory. This is because of the formation of fructose resulting from hydrolysis, which is strongly levorotatory than the glucose. Thus, the change of optical rotation of sucrose solution from dextro to levo rotation upon hydrolysis is known as inversion and the mixture of glucose and fructose obtained is called **invert sugar.**

Sucrose do not reduce metallic ions (i.e. do not answer Benedict's and Barfoed's tests) and also do not form osazone with phenyl hydrazine. This is because the functional groups of constituent monosaccharides, glucose and fructose are in α 1,2 linkage and thus no free functional aldehyde or keto group in sucrose.

But prolonged boiling with phenyl hydrazine in acid medium will form osazone due to the reaction of products of hydrolysis of sucrose (glucose and fructose) with phenyl hydrazine and **not due to the reaction of intact sucrose molecules with phenyl hydrazine.**

Reactions of Disaccharides

Molisch Test

Principle: Response of the disaccharides: All the disaccharides that are experimented routinely give the positive reaction—***reddish violet ring** as this is a general test to detect the presence of carbohydrate.*

Procedure, Observation and Inference: Same as given under monosaccharides.

Benedict's Test

Procedure, Observation and Inference: Same as given under monosaccharides.

Response of the Disaccharides
Based on Benedict's test disaccharides are classified into:
- Reducing disaccharides, e.g. lactose, maltose.
  These disaccharides have a free carbonyl (keto/aldehyde) group, not involved in glycosidic linkage will reduce cupric ions in the alkaline medium as explained under monosaccharides, e.g. lactose, maltose.
- Nonreducing disaccharides, e.g. sucrose, trehalose.
  These are the disaccharides in which the functional groups of constituent monosaccharides are in linkage. In sucrose aldehyde group of glucose and keto group of fructose are in glycosidic linkage where as in trehalose (present in mushrooms) aldehyde groups of two constituent glucose residues are linked by glycosidic bond (α 1→1 linkage).

Barfoed's Test

Procedure, Observation, Inference and Principle: Same as given under monosaccharides.

Response of the disaccharides: Disaccharides will not reduce cupric ions in the weak acid medium within the prescribed keeping time of 2 minutes in the boiling water bath.

Application: Useful to differentiate monosaccharides from disaccharides.

Points to Ponder
- If the heating time is prolonged, disaccharides will also give a positive response to Barfoed's test.

- If the concentration of disaccharide solution is high, Barfoed's test tends to become positive.

Osazone Test

Procedure: Same as given under monosaccharides except for the period for which the reaction tube to be placed in the boiling water bath—it is 45 minutes for disaccharides.

Lactose gives a characteristic **yellow puff shaped lactosazone** crystals (Fig. 1.15).

Maltose: Individual crystals of maltosazone looks like a yellow colored petal when grouped looks like a **sun flower** (Fig. 1.16).

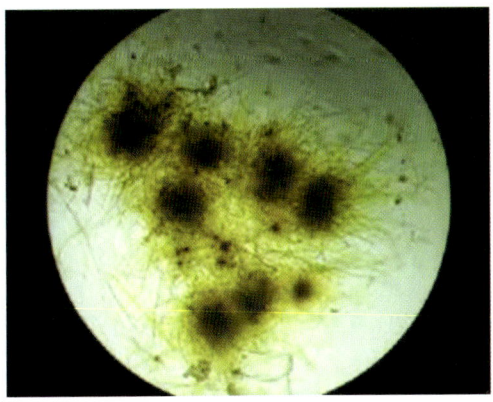

Figure 1.15 Lactosazone (Puff shaped)

Figure 1.16 Maltosazone (Petal shaped individual crystals and sunflower shaped when clustered)

Inference

Lactose → Puff shaped lactosazone crystals
Maltose → Petal shaped or sunflower shaped maltosazone crystals
Sucrose → Will not form osazone.

Principle: Reducing disaccharides with a reactive carbonyl group condense with phenyl hydrazine to form respective osazone crystals with characteristic shapes as detailed above.

Application: Useful to differentiate disaccharides.

Seliwanoff's Test

Principle: The disaccharide sucrose contains glucose and fructose. Fructose formed from sucrose upon acid hydrolysis by the HCl of Seliwanoff's reagent, is dehydrated by the acid HCl to form hydroxymethyl furfural which then condenses with the resorcinol of Seliwanoff's reagent to form a **red colored** complex.

Procedure: Same as given under monosaccharides.

Observation: Sucrose gives **bright red color** (Fig. 1.8) whereas lactose and maltose do not give red color.

Inference: Sucrose upon acid hydrolysis by the HCl in the Seliwanoff's reagent yields a keto sugar, fructose. Fructose being a keto sugar gives positive response to Seliwanoff's test as described under monosaccharides. Whereas lactose (galactose + glucose) and maltose (glucose + glucose) contain no keto sugar so cannot give positive response to this test upon acid hydrolysis by the HCl present in the Seliwanoff's reagent.

Rapid Furfural Test

Principle: The disaccharide sucrose contains glucose and fructose. Fructose formed from sucrose upon acid hydrolysis by the HCl, is dehydrated by the same HCl to form hydroxymethyl furfural which then condenses with the α-naphthol of Molisch reagent to form a **violet colored** complex.

Procedure: Same as given under monosaccharides.

Observation: **Sucrose** gives **violet color** (Fig. 1.6) whereas **lactose and maltose** do not give violet color.

Inference: Sucrose upon acid hydrolysis by the HCl added in the test yields a keto sugar fructose. Fructose being a keto sugar gives positive response to rapid furfural test as described under monosaccharides. Whereas lactose (galactose + glucose) and maltose (glucose + glucose) contain no keto sugar so cannot give positive response to this test.

Foulger's Test

Principle: The disaccharide sucrose contains glucose and fructose. Fructose formed from sucrose upon acid hydrolysis by the 40% H_2SO_4 is dehydrated by the same acid to form hydroxymethyl furfural which then condenses stannous chloride and urea to form deep blue color.

Procedure: Same as given under monosaccharides.

Observation : **Sucrose** gives **deep blue** (Fig. 1.9) whereas **lactose and maltose** do not give violet color.

Inference: Sucrose upon acid hydrolysis by the 40% H_2SO_4 in the reagent yields a keto sugar fructose. Fructose being a keto sugar gives positive response to Foulger's test as described under monosaccharides. Whereas lactose (galactose + glucose) and maltose (glucose + glucose) contain no keto sugar and cannot give positive response to this test.

Specific Sucrose Test (Fig. 1.17)

Procedure: It is done in two steps.

Step—1 Hydrolysis
To 3 mL of sucrose solution add 1 drop of thymol blue indicator and one or two drops of dilute HCl to make the solution acidic as shown

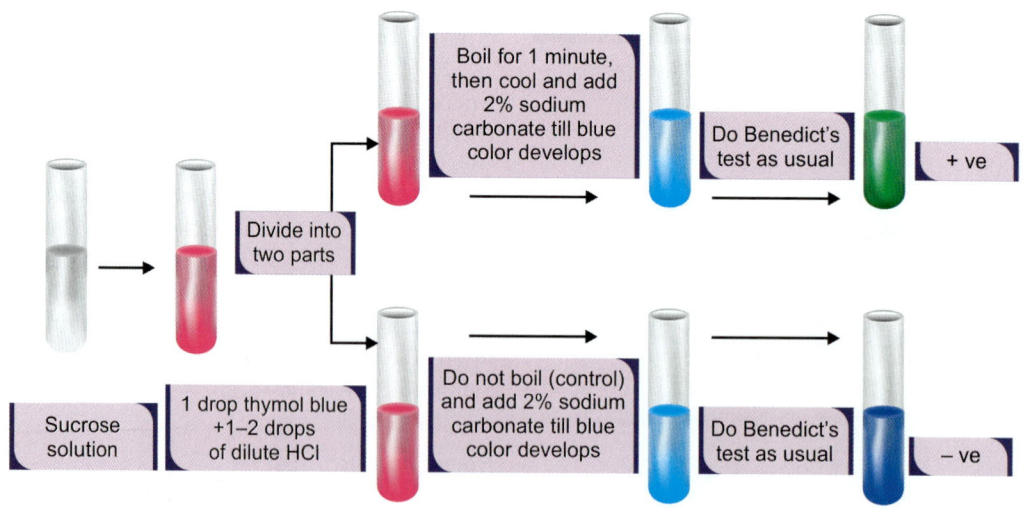

Figure 1.17 Specific sucrose test

by the development of pink color. Divide it into two equal parts. Boil one part for 1 minute and the other part is kept as control. Neutralize both parts by adding 20% sodium carbonate drop by drop until a blue color develops.

Step—2 Benedict's Test on Hydrolysate Obtained from Step 1
Perform Benedict's test with each portion.

Observation: Unboiled sucrose solution will not give a positive response to Benedict's test where as boiled portion gives a positive response.

Inference: Sucrose is hydrolyzed by HCl in the first step to form glucose and fructose and then the medium is neutralized by the 20% sodium carbonate.

In the second step, products of acid hydrolysis, reduce cupric ions to red cuprous oxide.

Precautions
- Avoid adding excess acid because it will dehydrate sugar to form furfural derivatives and that will interfere the test.
- Always remember to add alkali as per the test procedure since neutralization of acidic pH is needed for getting correct reaction in the second step.

Thymol blue indicator contains two components that work at acid **range (pH range 1.2–2.8; color change—red to yellow)** and at alkaline **range (pH range 8.0–9.6; color change—yellow to blue)**.

Alkali Destruction Test

Procedure: To 3 mL sugar solution add 0.5 mL of 40% NaOH. Mix well and keep in a boiling water bath for exactly 3 minutes.

Perform Benedict's test with the above solution.

Observation: The solution becomes yellow upon treatment with NaOH in boiling water bath for three minutes.

With Benedicts test, disaccharides give positive response.

Inference: Disaccharides escape complete destruction and give a +ve response to Benedict's test. But monosaccharides will give negative response due to complete destruction by the concentrated alkali.

REACTIONS OF POLYSACCHARIDES

Introduction

The polysaccharides are complex carbohydrates of high molecular weight, which on hydrolysis yields monosaccharides or products related to monosaccharides. The various polysaccharides differ from one another with respect to their constituent monosaccharide composition, molecular weight and other structural features.

In all types, the linkage between the monosaccharide units is the glycosidic bond. This may be α or β which join the respective units through $1 \rightarrow 2$, $1 \rightarrow 3$, $1 \rightarrow 4$ or $1 \rightarrow 4$ linkages in the linear sequence or at branch points in the polymer.

Polysaccharides are **classified** based on the type of monosaccharide units present in them.
- **Homopolysaccharide:** It contains only one type of monosaccharide, e.g. starch, glycogen.
- **Heteropolysaccharide:** It contains more than one type of monosaccharide units, e.g. Glycosaminoglycans (heparin, hyaluronic acid).

We will discuss the reactions of starch in this chapter in order to understand the chemical properties of polysaccharides in general.

REACTIONS OF STARCH

Molisch Test

Procedure, Observation and Inference: Same as given under monosaccharides.

Principle: The test is answered by all furfural yielding substances and hence all the carbohydrates.

Iodine Test

Procedure: To 2–3 ml of starch solution add 2 drops of dilute (0.05 N) iodine solution. Observe the changes on heating and on subsequent cooling.

Observation: Deep blue color appears which then disappears on heating and then reappears on cooling (Fig. 1.18).

Chemistry and Inference

Starch forms an adsorption complex with iodine to give a blue color. The blue color disappears on heating due to the breaking of the Iodine starch adsorption complex and appears on cooling due to reformation of the adsorption complex.

Benedict's Test

Procedure: Same as given with monosaccharides.

Observation: No colored precipitate.

Inference: Starch is a nonreducing carbohydrate.

Starch Hydrolysis Test

Procedure: Take 25 mL of starch solution in a beaker. Add 10 drops of concentrated HCl and boil gently. At the end of each minute, transfer a drop (using glass tube) of the solution on to a plate for doing the iodine test and 3 drops to 5 mL of Benedicts solution (Set tubes containing 5 mL of Benedict's reagent in series). Continue until the iodine test becomes negative. Then place the tubes for the Benedict's test in the boiling water bath for 3 minutes.

Observation: See Table 1.3.

Chemistry and Inference: Starch upon hydrolysis by HCl gives the following products. Starch → Soluble starch → Amylodextrins → Erythrodextrins → Achrodextrins → Maltose → Glucose. When the hydrolytic stage reaches to the level of formation of maltose and glucose iodine test becomes negative and Benedict's test becomes positive.

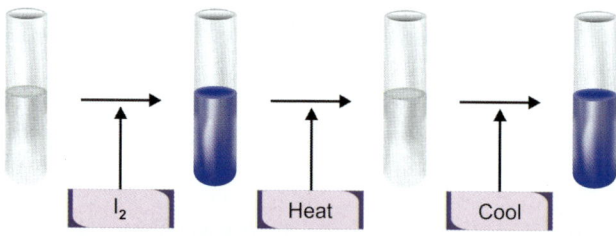

Figure 1.18 Iodine test

TABLE 1.3 Response of starch hydrolysis test

Time in minutes	Color with I_2	Benedict's test	Hydrolysis	Product
1	Blue	Blue	No reduction	Starch
5	Violet	Green	Reduction starts	Amylodextrins
8	Reddish violet	Red	Initiation of reduction	Amylo and erythrodextrins
12	No color	Red	Partial reduction	Achrodextrins
20	No color	Red	Complete reduction	Glucose

IDENTIFICATION OF UNKNOWN CARBOHYDRATES

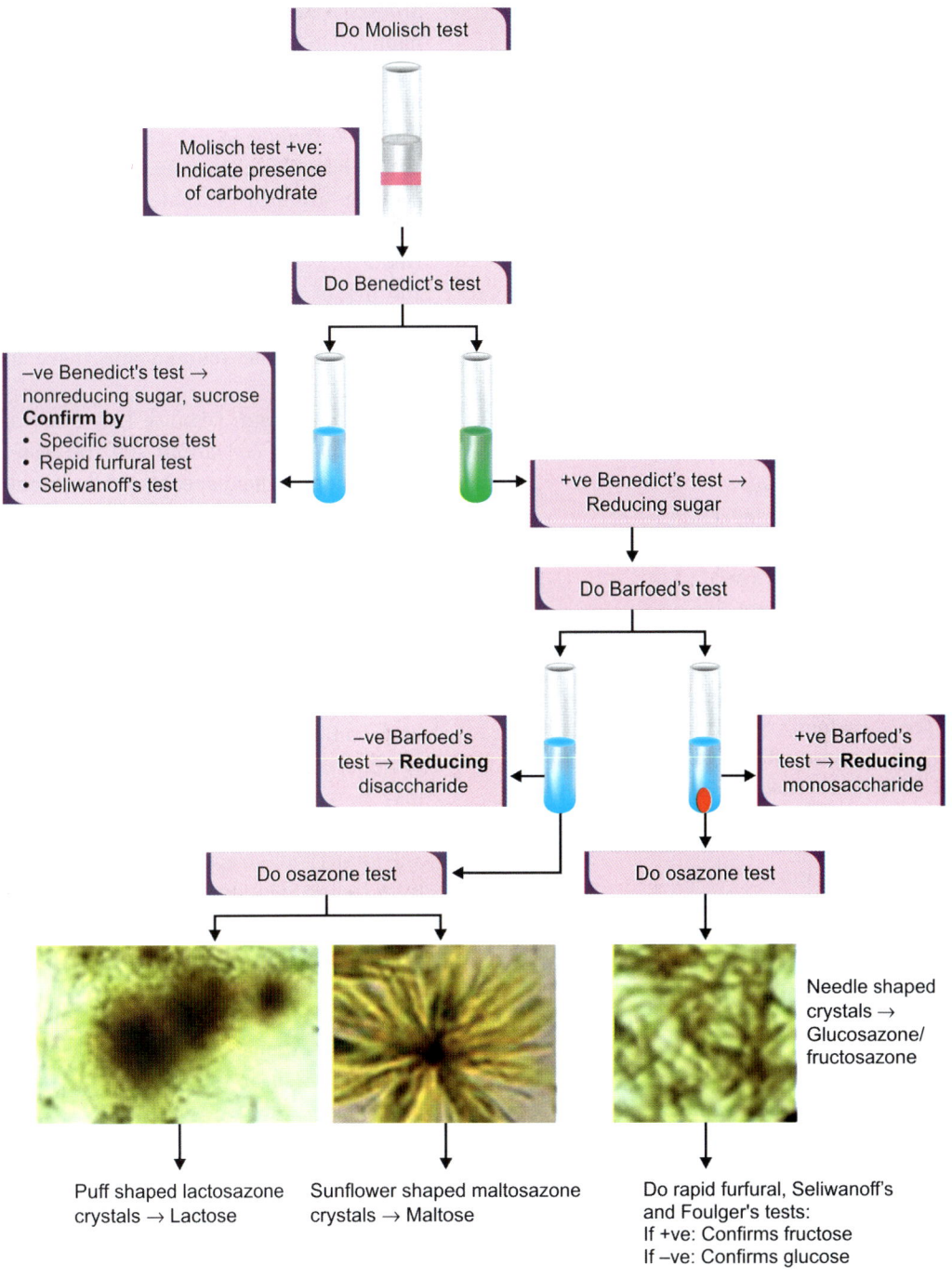

QUESTIONS

1. Name the following:
   a. General test for detecting carbohydrates
   b. Reduction test for monosaccharides
   c. Sugars giving positive response for rapid furfural test and Seliwanoff's test
   d. The disaccharide yielding puff shaped osazone crystals
   e. Tests based on reduction property of sugars
   f. The test used to detect sugar in urine
   g. Reducing disaccharides
   h. Nonreducing disaccharides
2. Give the principle of the following tests:
   a. Molisch test
   b. Benedict's test
   c. Barfoed's test
   d. Osazone test
   e. Iodine test for starch
   f. Rapid furfural test
   g. Seliwanoff's test
   h. Foulger's test
3. Give the ingredients of following reagents:
   a. Molisch's reagent
   b. Benedict's reagent
   c. Barfoed's reagent
   d. Seliwanoff's reagent
4. Benedict's test is described as a semi-quantitative test. Explain.
5. Barfoed's test is not suitable for testing glucose in urine. Why?
6. Give the differences between Benedict's and Barfoed's tests.
7. Why do glucose, mannose and fructose give similar osazone crystals?
8. Sucrose does not form osazone crystals with osazone test. Why?
9. Make a scheme for the detection of an unknown carbohydrate solution.

REAGENT PREPARATION

- **Molisch's Reagent:** Dissolve 5 g of α-naphthol in 100 mL of 95% of alcohol.
- **Benedict's Qualitative Reagent:** Heat to dissolve 173 g sodium citrate and 100 g sodium carbonate in about 800 mL of water in a conical flask. Transfer to a graduated cylinder through a folded filter paper placed in a funnel or beaker of 1 L capacity. Dissolve 17.3 g copper sulfate in about 100 mL of water. Add the copper sulfate solution slowly with constant stirring to the carbonate—citrate solution and make up to 1 L.
- **Barfoed's Reagent:** Dissolve 13.3 g neutral copper acetate crystals in 200 mL water. Pass through a filter paper placed in a funnel to remove the particles if present to another graduated beaker. Then add 1.8 mL glacial acetic acid.
- **Seliwanoff's Reagent:** Dissolve 0.05 g resorcinol in 100 mL dilute HCl.
- **Phenylhydrazine Mixture:** Mix 2 parts phenyl hydrazine hydrochloride and 3 parts sodium acetate by weight thoroughly in a mortar. (Mixture with longer shelf life may be prepared by using equal weights of phenyl hydrazine hydrochloride and anhydrous sodium acetate).
- **0.1 N Iodine Solution:** Dissolve 1.27 g iodine and 3 g pure KI (potassium iodide) crystals in 100 mL distilled water. Dilute 1:10 in distilled water before use.
- **Glucose, Fructose, Lactose, Maltose, Sucrose Solutions:** 1% solutions weigh 1 g of respective sugars and dissolve in 100 mL of water.

CHAPTER 2

Reactions of Proteins

INTRODUCTION

Proteins are the most abundant organic molecules (carbon containing) in the living system. They offer structural and dynamic functions. They are polymers of amino acids linked by covalent peptide bonds. Proteins ingested, undergo digestion and absorbed as amino acids into the portal vein and reaches liver and then to other tissues.

They are used mainly for protein synthesis as dictated by the genes of respective tissues (differential expression). Some amino acids undergo specific metabolic reactions to produce specialized compounds, e.g. epinephrine and norepinephrine formed from tyrosine, serotonin from tryptophan.

Housekeeping proteins like aldolase have longer half-life whereas *regulatory proteins* like HMG CoA reductase has shorter half-lives. After their life span proteins are catabolized to release nitrogen, which ultimately converted into urea and excreted in urine whereas the carbon skeletons are utilized for other purposes like gluconeogenesis.

Proteins are classified into **fibrous**, which offer mainly structural function, e.g. fibrinogen, troponin, collagen, myosin and **globular** proteins which offer mainly dynamic functions, e.g. Hb, enzymes, peptide hormones, enzymes, plasma proteins. Proteins are present in all types of body fluids.

During routine analytical laboratory work, two types of reactions are done.
1. **Precipitation reactions**
2. **Color reactions**

PRECIPITATION REACTIONS OF PROTEINS

Proteins have to be precipitated for different purposes during routine laboratory works. Two such situations are described below:
1. For identification and estimation, e.g. Proteins are excreted in urine in various forms of kidney dysfunction. According to the degree of kidney damage different proteins are excreted in urine. In the early stages, low molecular weight albumin is excreted. As the disease progresses high molecular weight globulin starts excreted.
2. For the analysis of other compounds in the specimen, proteins are first precipitated out.

Proteins form emulsoid colloidal solutions (colloid solutions are formed by particles with a diameter ranging from 1 μm to 200 μm).

Emulsoids (here proteins) in general possess **two stability factors—charge and water of hydration** either of these prevent aggregation and precipitation of proteins. The electrical charges carried by the proteins may be changed in sign or magnitude by changing the acidity or alkalinity of the solution causing them to precipitate. The inorganic salts like ammonium sulfate act as dehydrating agent, thereby removing the shell of hydration of the proteins. The dehydration is also carried out by organic solvents like alcohol and ether.

Precipitation by Salts (Fig. 2.1)

Inorganic salts when added to the protein solutions, water of hydration which form around the protein molecules is removed causing aggregation of protein molecules leading to their precipitation. Proteins are lyophilic colloids as they have much affinity for the dispersion medium.

Half Saturation Test with Saturated Ammonium Sulfate Solution

Procedure: To 3 mL of protein solution add an equal volume of saturated ammonium sulfate solution. Mix and allow to stand for 5 minutes. Filter (for this take a round filter paper of 5 cm radius and fold it to form a cone and place in a funnel). Then place this funnel with the filter paper over a test tube and pour the contents of the tube through funnel. Perform biuret test with the filtrate using an equal volume of **40% sodium hydroxide** and 2 drops of 1% $CuSO_4$.

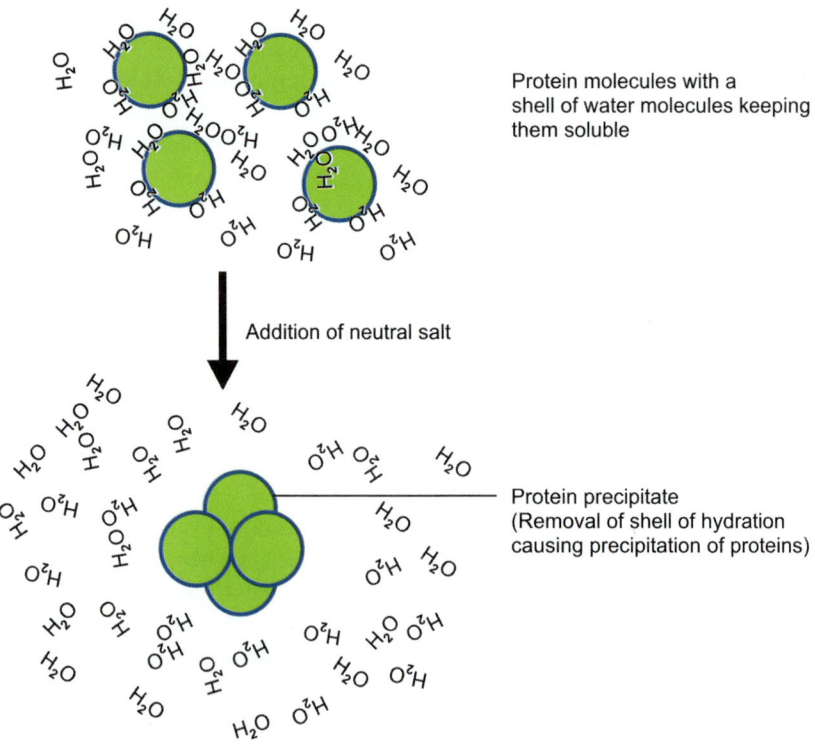

Figure 2.1 Principle of precipitation of proteins by salting out

Observation: Upon doing Biuret test with the filtrate, violet color forms.

Inference: Albumin is not precipitated by half saturation with ammonium sulfate (But globulins will precipitate with half saturation).

Principle: The molecular weight of the albumin is much less than the globulin, so albumin is not precipitated by half saturation (Fig. 2.2) whereas high molecular weight globulins are precipitated by saturated ammonium sulfate solution.

Points to Ponder

Use **40% sodium hydroxide for doing biuret test.**

In the routine biuret test 5% sodium hydroxide is used. But here the filtrate contains ammonium sulfate. Ammonium ions form a deep blue cuprammonium ion [Cu $(NH_3)_4^{++}$], which mask the violet color of biuret test. To avoid this 40% NaOH is used.

Full Saturation Test with Ammonium Sulfate Crystals

Procedure: To 5 mL of protein solution, keep on adding ammonium sulfate crystals while shaking the tube well, till a few crystals remain at the bottom of the test tube. Filter. For this, take a round filter paper of 5 cm radius and fold it to form a cone so as to fit it into a funnel. Then place this funnel over a test-tube and pour the contents of the test tube through the funnel. Perform biuret test with the filtrate using an equal volume of 40% sodium hydroxide and 2 drops of 1% $CuSO_4$.

Observation: Upon doing biuret test with the filtrate no purple or violet color develops (Fig. 2.2).

Inference: The protein (e.g. Albumin) is completely precipitated by full saturation with ammonium sulfate. Upon filtration no protein passes into the filtrate and hence not detected by the biuret test.

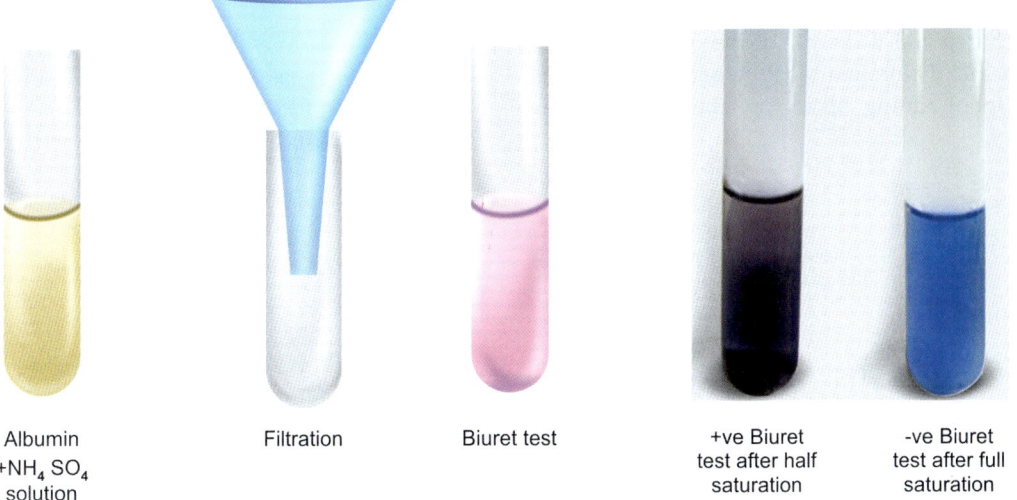

Albumin +NH₄ SO₄ solution | Filtration | Biuret test | +ve Biuret test after half saturation | −ve Biuret test after full saturation

Figure 2.2 Half saturation and full saturation tests of albumin

Principle: Neutral salt (e.g. ammonium sulfate) precipitate proteins by **salting out** which involves the removal of the shell of hydration causing precipitation of proteins. Higher the molecular weight lesser will be salt required for the precipitation. Globulins have much higher molecular weight than albumin so that globulins require only saturated solution whereas albumin require addition of salt for complete precipitation.

Precipitation by Heavy Metals (Fig. 2.4)

Precipitation by 10% Lead Acetate

Procedure: Take 3 mL protein solution; add 2 drops of 5% NaOH. Mix well and add 2 mL 10% lead acetate solution.

Observation: White precipitate forms.

Inference: Proteins are precipitated by positively charged lead ions.

Principle (Fig. 2.3): The isoelectric point of a protein is that pH at which the net charge on the protein is zero. If the pH of the medium is made alkaline the proteins acquire net negative charge and if the pH of the medium is made acidic the proteins acquire net positive charge. Upon **adding alkali** proteins gain **negative** charge and they form ionic bond with **positively charged metal ions** leading to precipitation of proteins.

Precipitation by 10% CuSO$_4$ Solution

Procedure: Take 3 mL of protein solution; add 2 drops of 5% NaOH. Mix well and add 10% CuSO$_4$ solution.

Observation: A light blue precipitate forms.

Inference: Proteins are precipitated by positively charged copper ions.

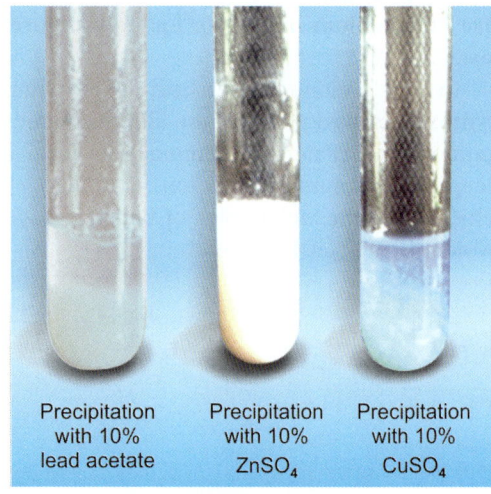

Precipitation with 10% lead acetate | Precipitation with 10% ZnSO$_4$ | Precipitation with 10% CuSO$_4$

Figure 2.4 Precipitation reactions of proteins by different metal ions

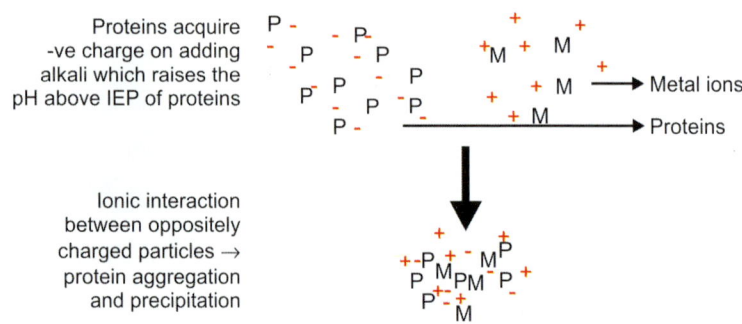

Figure 2.3 Mechanism of precipitation of proteins by metal ions in alkaline medium

Principle: The same as that of precipitation by lead acetate.

Precipitation by 10% ZnSO$_4$ Solution

Procedure: Take 3 mL of protein solution; add 2 drops of 5% NaOH. Mix well and add 2 mL of 10% ZnSo$_4$ solution.

Observation: An intense white precipitate forms.

Inference: Proteins are precipitated by positively charged zinc ions.

Principle: Similar to that of precipitation by lead acetate.

Precipitation by Anionic Reagents (Alkaloids) (Fig. 2.5)

Precipitation by Metaphosphoric Acid

Procedure: Take 3 mL of protein solution in a test tube and add a few drops of metaphosphoric acid.

Observation: A white precipitate forms.

Inference: Metaphosphoric acid in solution forms acid anion. Proteins become positively charged. Hence, positively charged protein ions and negatively charged acid anions derived from metaphosphoric acid combine to form insoluble complex leading to precipitation.

Principle: Alkaloids when dissolved, ionize and release protons leading to decrease in pH of the medium and formation of anions. Proteins in this acidic medium acquire positive charge and they complex with negatively charged anions (e.g. picrate from picric acid, sulfosalicylate from sulfosalicylic acid) derived from anionic reagents. These complexes are insoluble and thus precipitated.

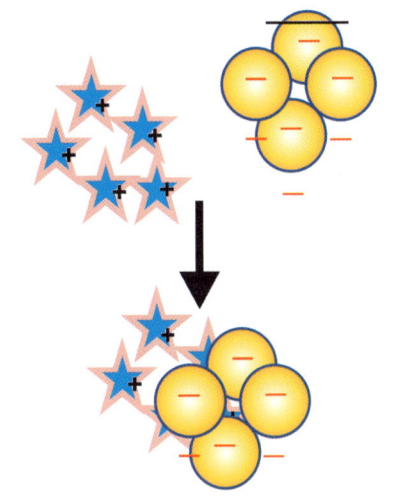

Figure 2.5 Mechanism of precipitation of proteins by anionic reagents

Precipitation by Organic Solvents (Fig. 2.6)

Precipitation by Ethanol

Procedure: Add 2 mL of ethanol to 1 mL of protein solution taken in a test tube and mix well.

Observation: Cloudy precipitate.

Inference: Proteins are precipitated due to removal of water of hydration by ethanol.

Precipitation by Heat

Heat Coagulation Test (Heat and Acetic Acid Test)

Procedure: Take a test tube and fill protein (albumin) solution up to two-thirds. Heat the

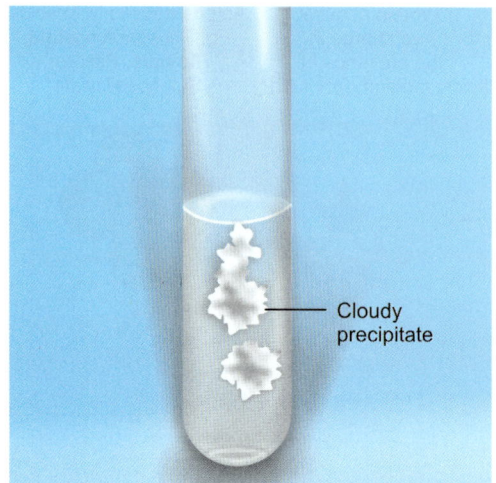

Figure 2.6 Precipitation by organic solvents

Figure 2.7 Precipitation by heating and influence of pH

upper 1/3rd of the column of protein solution. Note, whether any precipitate has appeared. Irrespective of the presence or absence of the appearance of the precipitate, add 2% acetic acid drop by drop. Note, whether the precipitate formed earlier (if any) becomes intensified or appears upon adding acetic acid.

Observation: White coagulum formed on initial heating intensifies on adding acetic acid (Fig. 2.7).

Inference: Albumin is denatured by heating and is precipitated by acetic acid.

Principle: Heating causes denaturation of albumin. Disruption of secondary, tertiary, quaternary structures maintained by noncovalent forces cause denaturation. Noncovalent forces include hydrogen bonds, ionic interactions, Van der Waal's forces and hydrophobic interactions. Aggregation of denatured protein is referred to as coagulum. Denaturation may be reversible in some cases (not always). But coagulation is always irreversible. Addition of acetic acid lowers the pH of the medium towards the isoelectric pH (pI) of the albumin (**IEP of different proteins: Human albumin: 4.7; egg albumin 4.9; human globulin 6.4; casein 4.6**). At pI, proteins are least soluble and hence the denatured proteins are precipitated upon adding acetic acid.

Precipitation by Strong Mineral Acids

Procedure: Take 2 mL of protein solution in a test tube. Add concentrated HNO_3 or concentrated HCl along the sides of the test tube slowly.

Observation: White ring forms.

Inference: Albumins as well as globulins are precipitated by strong mineral acids.

Principle: Strong acids causes denaturation and precipitation of proteins.

Points to Ponder: Precipitation by HNO_3 is named as **Heller's test**. It is used as a test for detecting protein in urine or other body fluids.

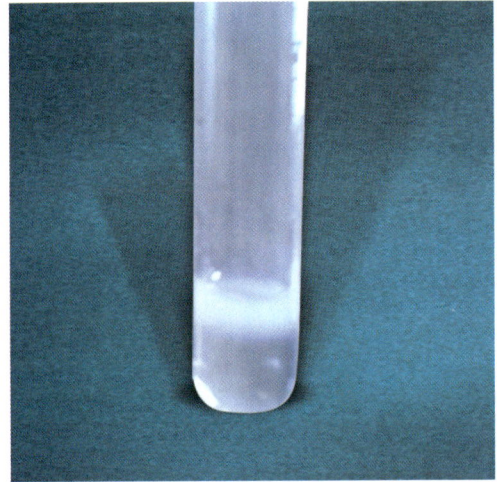

Figure 2.8 Heller's test

COLOR REACTIONS OF PROTEINS AND AMINO ACIDS

Proteins react with a variety of reagents to form colored products because of their constituent peptide bonds and amino acids. These reactions are useful for quantitative and qualitative studies of proteins. By quantitative studies the concentration of the proteins are estimated. Qualitative studies help to know the presence of proteins or specific amino acids present in the protein. They are useful mainly in the following situations:

- **For the diagnosis of aminoacidurias:** Individual amino acids undergo unique catabolic pathways and the deficiency of any enzyme of these pathways lead to accumulation of compounds proximal to the defective step causing disorders called aminoacidurias. For instance phenylketonuria due to phenylalanine hydroxylase deficiency causes elevated blood levels of phenylalanine in the blood and urine. Study of aminoaciduria, needs identification of specific amino acid in the body fluids. Study of color reactions of amino acids are useful in the screening of aminoacidurias.

- **For the nutritional assessment:** Out of twenty primary or standard amino acids, only eight are essential and the rest of the twelve amino acids are nonessential for adults. Those proteins containing all the essential amino acids are considered to be good quality proteins, e.g. egg albumin. Hence, for the nutritional assessment of proteins also, the study of reactions of amino acids is helpful.

- **To detect the presence of proteins or amino acids in biological fluids or in fluids with unknown composition:** This chapter deals with different color reactions of amino acids. The color reactions are due to the reaction between constituent radical or group of the amino acids and the chemical reagents used in the test. Amino acid composition of different proteins is different. Depending on the nature of amino acids contained in a protein, the response and the intensity of the color reactions varies.

Biuret Test

Procedure: To 2–3 mL of protein solution add an equal volume of 10% sodium hydroxide solution, mix thoroughly. Then add a 0.5%

Figure 2.9 Biuret test

copper sulfate solution drop by drop, mixing between drops until a violet color is obtained.

Observation: Violet color develops.

Inference: The biuret reaction is given by substances which contain two carbamyl groups (-CONH$_2$) joined either directly together or through a single atom of nitrogen or carbon. Positive reaction indicates that the given protein solution contains at least two peptide bonds.

Chemistry of the reaction: The biuret test is given by those substances containing two carbamyl groups (-CONH$_2$) joined either directly or by a single nitrogen or carbon atom. The purplish violet color is due to the formation of a copper coordination complex (Fig. 2.10).

The molecule should have a minimum of two peptide bonds to give copper coordination complex that impart **violet** color to test mixture. *This reaction is first carried out with the compound biuret formed by the condensation of 2 molecules of urea upon heating. This compound contains two peptide bonds as shown below* (Fig. 2.11).

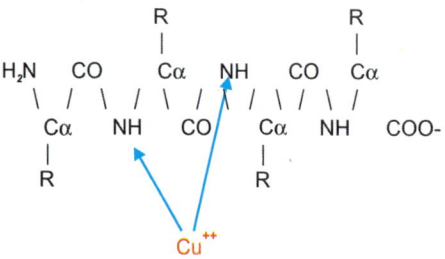

Figure 2.10 Copper coordination complex formed in biuret test

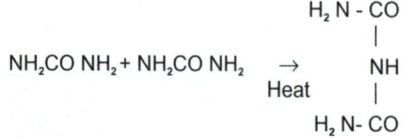

Figure 2.11 The compound biuret

Biuret (a nonprotein, formed *from urea on heating; biuret formed gives violet color with copper sulfate solution in the alkaline medium*).

Proteins give **violet color** with biuret test since there are several pairs of CONH groups in the molecule (Fig. 2.12).

Points to Ponder

- If too much copper sulfate solution is added, blue colored copper hydroxide will be formed and that will mask the violet color.
- If magnesium sulfate is present in the test solution it forms magnesium hydroxide and interferes with the test.
- If more ammonium sulfate is present excess alkali has to be used.

Ninhydrin Test

Procedure: To 1 mL of protein solution (pH must be between 5 and 7) in a test tube add 2–3 drops of freshly prepared 0.1% ninhydrin (triketohydrindene hydrate) solution. Heat the solution to boil for 2 minutes and allow to cool.

Observation: Purple (violet) color with α-amino acids and yellow with imino acid (Proline).

Inference: Purple (violet) color is due to the formation of a complex called Ruhemann's purple formed between N-terminal nitrogen and ninhydrin.

Chemistry of ninhydrin reaction: Ninhydrin oxidizes amino acid to form aldehyde, with the liberation of one molecule of ammonia and CO$_2$ and itself converted to hydrindantin.

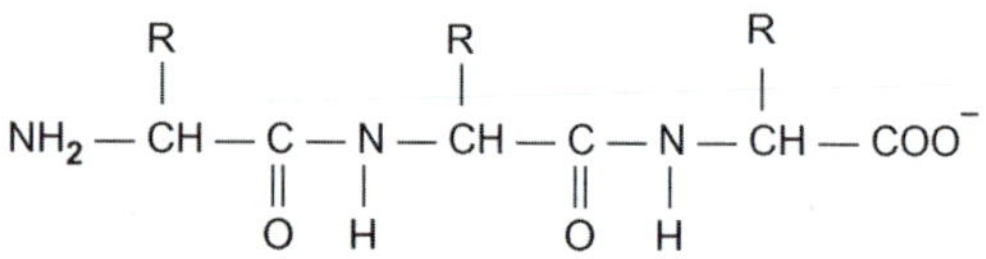

$$NH_2 - CH(R) - \underset{\underset{O}{\|}}{C} - N(H) - CH(R) - \underset{\underset{O}{\|}}{C} - N(H) - CH(R) - COO^-$$

Figure 2.12 CONH groups in a peptide

The hydrindantin thus formed complex with another molecule of ninhydrin and a molecule of ammonia to form a complex called **Ruhemann's purple**. **Proteins** give a **faint blue color**. In the case of proteins, amino terminal nitrogen participate in the action.

Application: Staining of amino acids in paper chromatography.

SPECIFIC COLOR REACTIONS USED TO IDENTIFY THE SIDE CHAIN (R) GROUPS OF AMINO ACIDS

Xanthoproteic Reaction (Figs 2.13 and 2.14)

Procedure: Add 1 mL of concentrated nitric acid to 2–3 mL of test protein solution. Heat to boil. Cool and pour half of the solution into another tube. One tube is kept as control and the other as test, so as to understand the development of even a faint color. To 'test' tube add 40% NaOH or liquor ammonia (ammonium hydroxide) in excess and compare with the control.

Observation: A white precipitate forms on adding nitric acid which on heating turns yellow and then dissolves to impart yellow color to the solution. Upon adding alkali the color deepens to attain **orange** color.

Interpretation: Addition of nitric acid causes denaturation of proteins to get white precipitate. Yellow color due to nitration of benzene ring

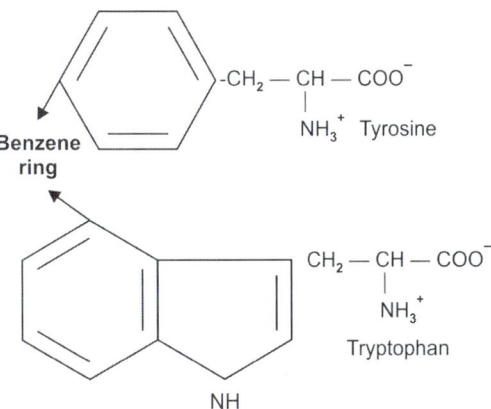

Figure 2.14 Tyrosine and tryptophan

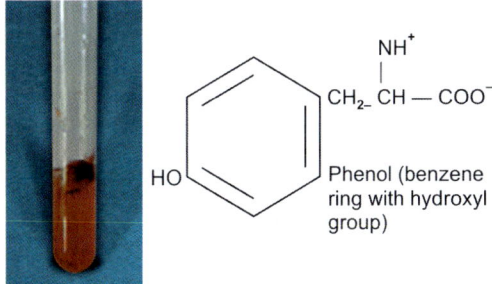

Figure 2.15 Millon's test

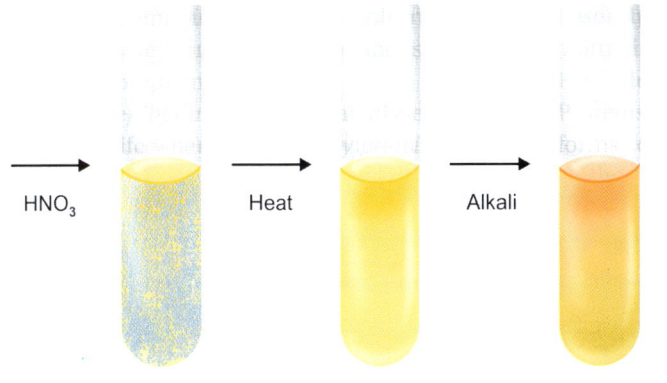

Figure 2.13 Xanthoproteic test

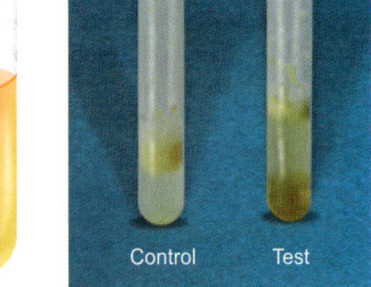

of amino acids—**tryptophan and tyrosine**. Addition of alkali increases the ionization of compounds hence the color deepens to get final orange color (Fig. 2.14).

Points to Ponder

This test cannot be employed for urine testing due to the final color of the test and the natural color of urine are similar.

Millon's Test

Procedure: To 2 mL of protein solution in a test tube, add 2 mL of 10% mercuric sulfate ($HgSO_4$) in 10% sulfuric acid (Millon's reagent). Boil for 30 seconds. A precipitate may form at this stage. Add a few drops of 1% $NaNO_2$ and gently warm.

Observation: Red precipitate forms and the solution turns **red**. Amino acid solutions gives red color without a precipitate (Fig. 2.15).

Inference: The given protein contains the amino acid tyrosine with a phenolic radical → positive Millon's test.

Principle: The protein precipitated by mercuric sulfate in acidic medium to form mercury—protein complex (metallo protein complex). Nitrous acid is formed by the reaction between sodium nitrite and sulfuric acid. This nitrous acid causes nitration of phenolic groups of tyrosine. Warming enhances nitration process and intensifies the color.

Aldehyde Test (Glyoxylic Acid, Hopkins-Cole Reaction)

Procedure: Take 2–3 mL of protein test solution, add 2 drops of 1/500 formaldehyde (HCHO) and 1 drop of 10% mercuric sulfate in sulfuric acid (Millon's reagent). Mix well. Add 3 mL of concentrated sulfuric acid through the sides of the test tube.

Observation: A **purple (violet)** ring develops at the junction of two layers (Fig. 2.16).

Inference: The **purple color** is due to the indole ring of the amino acid tryptophan.

Principle: Mercuric sulfate in sulfuric acid act as an oxidizing agent and it oxidizes the indole ring of tryptophan. Then formaldehyde react with the oxidized indole ring to form purple colored complex.

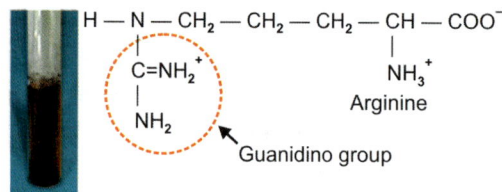

Figure 2.17 Sakaguchi's test and amino acid arginine

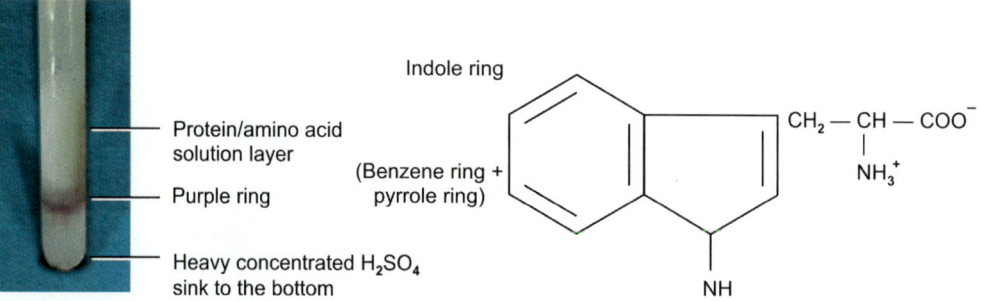

Figure 2.16 Aldehyde test and amino acid tryptophan

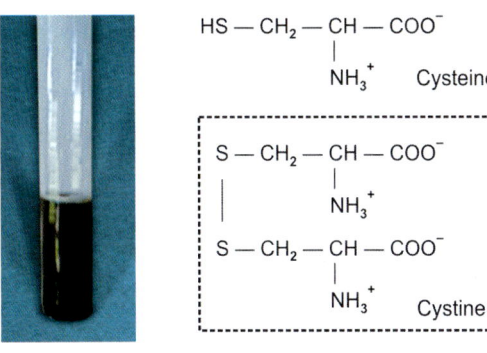

Figure 2.18 Sulfur test and amino acids cysteine, cystine and methionine

Sakaguchi's Test

Procedure: Add 5 drops of 5% sodium hydroxide to 5 mL protein solution. Shake well. Add 2–4 drops of Molisch reagent and add 2 mL freshly prepared bromine water.

Observation: A **bright red** (Fig. 2.17) color develops.

Inference: The given protein contains the amino acid arginine with guanidino group.

Principle: Molisch reagent is α-naphthol in alcohol. Sodium hydroxide provides alkaline pH. At the alkaline pH guanidino group of arginine combines with α-naphthol and hypobromite, to form bright red color.

Sulfur Test (Lead Blackening Test)

Procedure: To 3 mL of protein solution add 3 mL of 40% NaOH and boil for 3 minutes. Cool, add 1 mL of lead acetate solution.

Observation: Solution turns **dark brown** (Fig. 2.18).

Inference: This test is answered by "S" containing amino acids—cysteine and cystine but not methionine because of the position of S in the thioether linkage.

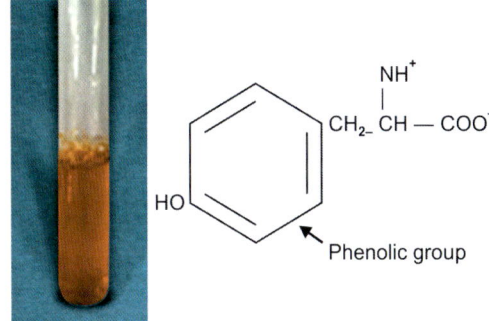

Figure 2.19 Pauly's test—Orange red (Tyr)

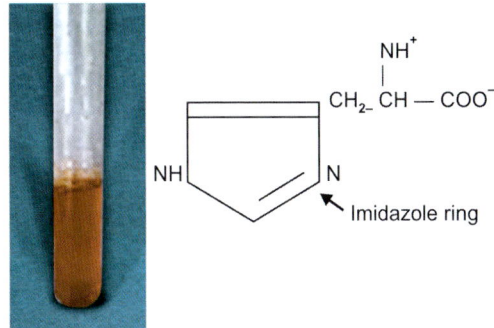

Figure 2.20 Pauly's test—Cherry red (His)

Principle: Upon boiling with strong alkali the organic sulfur in the cystine and cysteine is converted into sulfide (here Na_2S). The sodium sulfide react with lead acetate to form black lead sulfide (PbS) and solution turns **brownish black**.

Points to Ponder: **Casein** and **gelatin** give a faintly positive reaction due to the deficiency of cysteine.

Pauly's Test (Figs 2.19 and 2.20)

Procedure: To 0.5 mL of 0.5% sulfanilic acid, add an equal volume of 0.5% freshly prepared sodium nitrite. Allow to stand for 1 minute and add 1 mL of protein solution. Mix well and add 1 mL of 10% Na_2CO_3 to make the solution alkaline.

Observation: **Cherry red** or **orange red** color may be observed.

Inference: Cherry red color indicates the presence or predominance of **histidine** and orange red color shows the presence or predominance of **tyrosine** in the solution.

Principle: Diazotized sulfanilic acid when complexes with imidazole ring of histidine

TABLE 2.1 Reactions of albumin

	Precipitation Reactions	
Reaction	Observation	Inference
1. **Isoelectric precipitation:** Take 10 mL of protein solution in a test tube. Add 2–3 drops of Chlorophenol red (pH range—5.0–6.6; color range—yellow to red). The purpose of adding the indicator is to get pH around 5.0. Look at the color change 2% Na_2CO_3 → 2% acetic acid → With chlorophenol red, the yellow color denotes pH either equal to 5 or less than 5. So even if a yellow color is observed the pH may not be 5, it may be much less than pH 5 also. In order to make sure of the required pH 5, add 2% Na_2CO_3 in drops until a pink color forms and then add 2% acetic acid in drops till the solution turns just becomes yellow. Suppose if it is red or pink upon adding chlorophenol red, it is indicating that the prevailing pH is either 6.6 or more than that. Add 2% acetic acid in drops till a yellow color just develops. Boil the above solution	White coagulum	At pI, albumin is denatured. Upon heating denatured protein aggregate to form visible change named coagulation

Contd...

Contd...

Reaction	Observation	Inference
2. **Heller's test:** Take 2 mL of protein solution in a test tube. Add 2 mL of concentrated HNO_3 or concentrated HCl along the sides of the test tube slowly	A white ring forms (Fig. 2.8)	Stratification of acid over albumin solution causes denaturation and precipitation of albumin at the point of contact
3. **Half saturation test:** To 5 mL of albumin solution add equal volume of saturated ammonium sulfate solution. Shake vigorously for 2 minutes. Keep it for 5 more minutes. Filter and collect the filtrate Perform biuret test → To 2 mL of the above filtrate taken in a test tube add 2 mL 40% NaOH and 1% $CuSO_4$ drop by drop	Violet color (Fig. 2.2)	Albumin being relatively small in size (MW 69000 Kda) is not completely precipitated by saturated solution of ammonium sulfate and hence goes into the filtrate → positive biuret reaction
4. **Full saturation test:** To 5 mL of albumin solution add ammonium sulfate crystals and shake well till some crystals remain at the bottom of the tube. Keep it for 5 minutes and filter. Collect the filtrate Do biuret test with the filtrate → To 2 mL of the above filtrate taken in a test tube add 2 mL 40% NaOH and 1% $CuSO_4$ drop by drop	No violet color (Fig. 2.2)	Albumin is completely precipitated by full saturation with ammonium sulfate crystals. Hence, the filtrate do not contain albumin → negative biuret reaction
5. **Heat and acetic acid test /Heat coagulation test:** Take a test tube and fill protein (albumin) solution up to two-thirds. Heat the upper one-third portion of protein solution column. Note whether any precipitate has appeared. Irrespective of the presence or absence of the appearance of the precipitate, add 2% acetic acid drop by drop. Note whether the precipitate formed earlier (if any) has intensified or appeared upon adding acetic acid	Precipitate formed on heating becomes denser on adding acetic acid (Fig. 2.7)	Heating caused coagulation of albumin and the addition of acetic acid lowered the pH of the medium towards the isoelectric pH (pI) of the albumin and enhanced the precipitation
Color Reactions		
1. **Biuret reaction:** (Fig. 2.9)	Violet color	Albumin contains more than 2 peptide bonds
2. **Xanthoproteic reaction:** (Fig. 2.13)	Deepening of yellow color	Albumin contains tryptophan and tyrosine. Benzene ring of these amino acids are giving the reaction
3. **Millon's test:** (Fig. 2.15)	Red color	Albumin contains the phenolic group containing amino acid, tyrosine
4. **Aldehyde test:** (Fig. 2.16)	Violet or purple ring at the junction of two liquids	Indole ring containing amino acid tryptophan is present in albumin
5. **Sakaguchi's test:** (Fig. 2.17)	Bright red color	Albumin contains the guanidino group containing amino acid arginine
6. **Sulfur test:** (Fig. 2.18)	Brownish black solution	Indicates the presence of cysteine in albumin
7. **Pauly's test:** (Figs 2.19 and 2.20)	Orange red	Indicates predominance of tyrosine in the given albumin solution

gives cherry red colored complex and when it complexes with phenolic group of tyrosine yields orange red color.

REACTIONS OF ALBUMIN (TABLE 2.1)

Albumins are compact roughly spherical in shape and have axial ratios **not** more than 3 (that is the ratio of their shortest to longest dimensions). Hence, albumins come under **globular proteins.** They have definite molecular weight. Albumins of interest are serum albumin of blood, lactalbumin of milk and ovalbumin of egg. It is also present in pulses. They are soluble in solute free water and coagulable on heating. They are not precipitated by half saturation with salts. Its isoelectric point (pI) is 4.55 – 4.9. Egg albumin is commonly employed in the laboratory to carry out experiments to study the properties of albumin in general. Human serum albumin has got a molecular weight of 69,000 Kda and its **pI is 4.7.**

REACTIONS OF CASEIN (TABLE 2.2)

Casein is the main protein present in the milk. It is a phosphoprotein and constitute a 1/3rd of proteins of human milk, 5/6th of proteins of cow's milk and 3/4th of proteins of goat's milk. Casein is secreted by mammary gland only.

TABLE 2.2 Reactions of casein		
Precipitation Reactions		
Reaction	Observation	Inference
1. **Isoelectric precipitation:** Take 4 mL of protein solution in a test tube. Add 2–3 drops of Bromocresol green (pH range 4.0–5.6; color range—yellow to blue). The purpose of adding the indicator is to get pH around 4.6. Look at the color change. 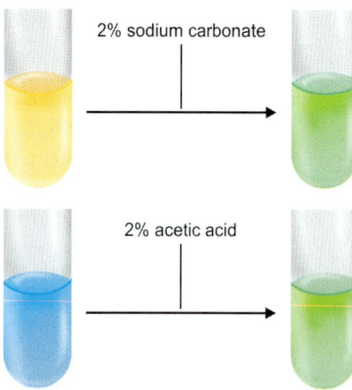 If it is yellow that means pH is ≤ 4.0. Then add 2% sodium carbonate drop by drop till the solution turns light green. (pH around 4.6) If it is blue that means pH is ≥ 5.6. Then add 2% acetic acid drop by drop till the solution turns light green. (pH around 4.6)	Precipitate seen	When the pH reaches 4.6 (pI of casein) casein precipitates

Contd...

Chapter 2 Reactions of Proteins

Contd...

Reaction	Observation	Inference
2. **Half saturation test:** To 5 mL of Casein solution add equal volume of saturated ammonium sulfate solution. Shake vigorously for 2 minutes. Keep it for 5 more minutes. Filter and collect the filtrate Perform biuret test. To 2 mL of the above filtrate taken in a test tube add 2 mL 40% NaOH and 1% $CuSO_4$ drop by drop	No violet color with biuret test	Casien is completely precipitated by saturated solution of ammonium sulfate. So the filtrate does not contain any casein as it is completely filtered off due to precipitation. Hence, the biuret test done with filtrate becomes negative
3. **Heat and acetic acid test/Heat coagulation test:** (given under reactions of albumin)	No coagulation	Casein not heat coagulable

Color Reactions: All the color reactions will be positive except the sulfur test. Sulfur test will be faintly positive because only 0.3 g of cysteine/cystine is present in 100 g of casein [whereas egg albumin contains about 2.5 g cysteine or cystine per 100 g]

Specific tests for casein		
1. **Neumann's test (detect organic phosphorous):** To 5 mL of casein solution add 0.5 mL of 40% NaOH. Heat for one minute and cool it by keeping in a rack. Add 0.5 mL of concentrated nitric acid. Add 1 mL of saturated ammonium molybdate solution	Canary yellow precipitate	Casein is digested by heating with sodium hydroxide thereby inorganic phosphorous (Pi) is released from it. Ammonium molybdate react with Pi in the acidic medium provided by the concentrated nitric acid to form **ammonium phospho molybdate** which is canary yellow in color

TABLE 2.3 Reactions of gelatin

Precipitation Reactions		
Reaction	Observation	Inference
1. **Half saturation test:** To 5 mL of gelatin solution add equal volume of saturated ammonium sulfate solution. Shake vigorously for 2 minutes. Keep it for 5 more minutes. Filter and collect the filtrate Perform biuret test: To 2 mL of the above filtrate taken in a test tube add 2 mL 40% NaOH and 1% $CuSO_4$ drop by drop	No violet color with biuret test	Gelatin is completely precipitated by saturated solution of ammonium sulfate. So the filtrate does not contain any gelatin as it is completely filtered. Hence, the biuret test done with the filtrate becomes negative

Color Reactions: Aldehyde test: - ve, Millon's test and Sulfur test: faintliy +ve - because tryptophan is absent in gelatin; Tyr and Cys present in very low amounts. All the other color reactions will be positive

Contains all the essential amino acids. It is less soluble and is made soluble at acid or alkaline pH and precipitate when the pH is brought to isoelectric point. (4.6). It is completely precipitated by half saturation with ammonium sulfate and it is not coagulated by heat. Casein act like a suspensoid, the particles of it flocculate when their charges are neutralized.

REACTIONS OF GELATIN (TABLE 2.3)

Gelatin is derived from fibrous protein collagen which is abundantly present in the ligaments, bones and teeth. Upon treating with boiling water and subsequent cooling, collagen give rise to gelatin which is free from carbohydrate moieties. It has low biological value since it is

IDENTIFICATION OF UNKNOWN PROTEINS

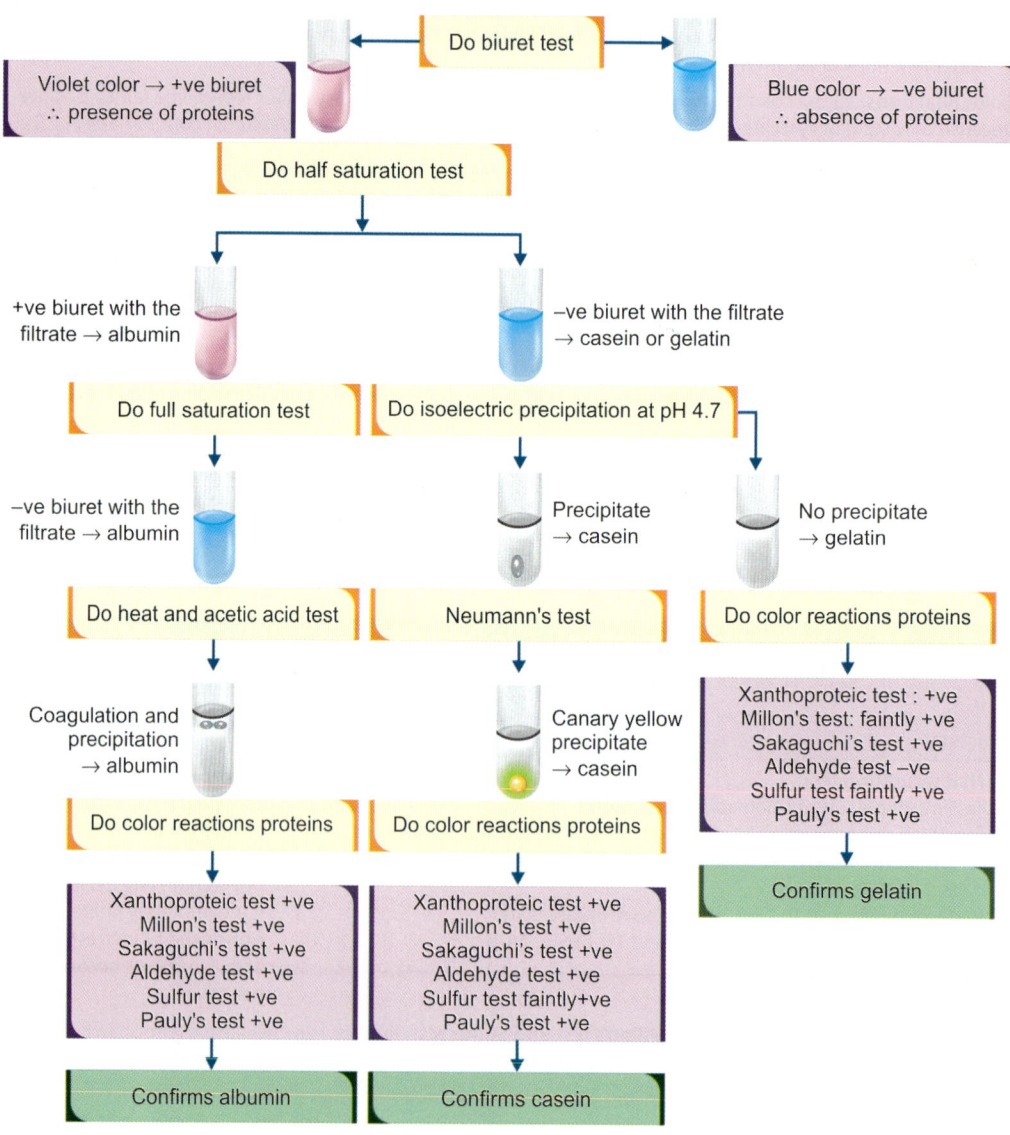

lacking the essential amino acid tryptophan and contains very low amounts of phenylalanine, tyrosine and cysteine. The **isoelectric point for gelatin is about pH 4.7.** But it is quite soluble in water even at its isoelectric point (IEP).

QUESTIONS

1. Name the following:
   a. General test for detecting proteins
   b. Test to detect peptide linkage
   c. A phosphoprotein
   d. Test to detect the presence of tryptophan
   e. Test to detect the presence of arginine
   f. Test to detect the presence of tyrosine
   g. Test to detect the presence of cysteine
   h. Test to detect the presence of cystine
   i. Color produced by α-amino acid and imino acid with ninhydrin
2. Give the principle of the following tests:
   a. Biuret test
   b. Ninhydrin test
   c. Half saturation test
   d. Full saturation test
   e. Isoelectric precipitation of proteins
   f. Sakaguchi's test
   g. Millon's test
   h. Xanthoproteic test
   i. Aldehyde test
3. Give brief answers:
   a. Denaturation vs coagulation
   b. Structural alterations of proteins in denaturation
   c. Mechanism of precipitation of proteins by salts like ammonium sulfate
   d. Mechanism of precipitation of proteins by alkaloids
   e. Mechanism of precipitation of proteins by alcohol
   f. Uses of color reactions in a diagnostic laboratory
   g. Applications of precipitation of proteins in the clinical chemistry laboratory
   h. Rationale of using egg white as a part of treatment of mercury poisoning
   i. What are the amino acids detected by Sulfur test?
   j. Why methionine is not responding to sulfur test?
   k. What is the reason for using 40% NaOH for doing biuret test with filtrate obtained after half and full saturation with ammonium sulfate?

REAGENT PREPARATION

- **1% Albumin Solution:** Dissolve 1 g of albumin in 100 mL of water. Ideally prepare on the day or on the day before experimentation and store in the refrigerator.
- **1% Casein Solution:** Dissolve 1 g of casein in 100 mL of 0.1 N NaOH. Ideally prepare on the day or on the day before experimentation and store in the refrigerator.
- **1% Gelatin Solution:** Dissolve 1 g of gelatin in 100 mL of water. Ideally prepare on the day or on the day before experimentation and store in the refrigerator.
- **Ammonium Molybdate Solution:** Dissolve 100 g molybdic acid in 144 mL of ammonium hydroxide (specific gravity 0.90) and 271 mL water. Slowly with constant stirring pour the solution into 489 mL of nitric acid (specific gravity 1.42) and 1148 mL water. Keep in a warm place for several days till a portion heated to 40°C deposits no yellow precipitate of ammonium phosphomolybdate. Decant the solution from any sediment and keep in glass stoppered bottles.
- **Lead Acetate Solution (10%):** Dissolve 10 g of lead acetate in 100 mL of distilled water.
- **Sodium Hydroxide 1N Solution:** Sodium hydroxide always contains sodium carbon-

ate as an impurity derived from CO_2 from air. Besides it is hygroscopic (absorb water from atmosphere). So, direct preparation of solution, will not give the concentration required for the purpose.

Molecular weight of NaOH = 23 + 16 + 1 = 40

Equivalent weight of NaOH = 40/1 = 40

Therefore, for making 1N NaOH solution, dissolve 100 g reagent grade sodium hydroxide pellets in 100 mL of distilled water in a flask and cap it and leave overnight at room temperature. The carbonate will settle out as an insoluble precipitate. Filter the solution through a sintered glass filter.

Pipette 75 mL and make up to 1 L into a 1000 mL volumetric flask and dilute to 1000 mL with distilled water. Mix well.

Prepare 1N oxalic acid by dissolving 63.035 g of pure crystalline oxalic acid in distilled water and make up to a volume of 1 L.

To make 1N NaOH—Take 10 mL of 1N oxalic acid in a conical flask and titrate against the NaOH solution prepared in the burette using Phenolphthalein as an indicator. (color range : colorless to red; pH range: 8-9.8).

If the titer value (Number milliliters of NaOH required to neutralize 10 mL of oxalic acid) is 9.5

Then the normality of NaOH solution,

= 1N × 10/ 9.5 = 1.05 N

To prepare 1N NaOH,

= 1000 mL × 1N/Normality of the solution

= 1000 mL × 1N/1.05 N = 952.4

The take 952.4 mL NaOH solution into a 1L measuring cylinder and dilute to 1000 mL by adding distilled water.

- **Sodium Hydroxide 0.1N Solution:** Dilute 100 mL of 1N NaOH to 1 liter with distilled water.
- **40% Sodium Hydroxide:** 1N NaOH solution is equivalent to 40%.
- **10% Copper Sulfate Solution:** Weigh 10 g copper sulfate into a few mL of distilled water taken in a 100 mL flask and dissolve and make up to 100 mL with distilled water.
- **0.5% Copper Sulfate Solution:** Weigh 0.5 g copper sulfate into a few mL of distilled water taken in a 100 mL flask and dissolve it and make up to 100 mL with distilled water.
- **10% Zinc Sulfate Solution:** Weigh 10 g zinc sulfate into a few mL of distilled water taken in a 100 mL flask and dissolve it and make up to 100 mL with distilled water.
- **10% Mercuric Sulfate in Sulfuric Acid:** Dissolve 10 g of mercuric sulfate in 100 mL of 10% sulfuric acid.
- **0.1% Ninhydrin Solution:** Dissolve 0.1 g of ninhydrin in 100 mL of acetone.
- **1% Sodium Nitrite ($NaNO_2$):** Dissolve 1 g sodium nitrite in 100 mL of water.
- **1/500 Formaldehyde:** Dissolve 1 mL of formaldehyde in 500 mL of distilled water.
- **Liquor Ammonia:** Available as such commercially.
- **0.5% Sulfanilic Acid:** Dissolve 0.5 g of sulfanilic acid in 100 mL of 2% HCl.
- **10% Sodium Carbonate:** Dissolve 10 g of sodium carbonate in 100 mL of distilled water.
- **2% Sodium Carbonate:** Dissolve 2 g of sodium carbonate in 100 mL of distilled water.
- **Chlorophenol Red: pH range 5-6.6; color range—yellow to red):** Add 0.1 g chlorophenol red and 4.8 mL of 0.05N NaOH to 250 mL of distilled water.
- **Bromocresol Green (pH range 3.8-5.4; color range—yellow to green):** Add 0.1 g Bromocresol green and 3.7 mL of 0.05N NaOH to 250 mL of distilled water.
- **Phenolphthalein (pH range 8.3-10; color range—colorless to red):** Dissolve 0.1 g in 100 mL of 50% ethanol.

CHAPTER 3

Reactions of Lipids

GENERAL REACTIONS OF LIPIDS

Introduction

Lipids are naturally occurring heterogeneous group of substances found in all vegetable and animal matter. They are insoluble in water and soluble in solvents like ether, chloroform, boiling alcohol and benzene. Lipids are esters of fatty acids or substances capable of forming such esters. To understand much about the different types of lipids classification of lipids is useful (Table 3.1). Plasma lipids are present as complexes with protein molecules which make the lipids soluble complexes and are called lipoproteins.

REACTIONS OF FATS AND FATTY ACIDS

Solubility Test

Procedure: Take 4 dry tubes and arrange them in a test tube stand. Add 2 mL each of **water,**

TABLE 3.1 Classification of lipids

Simple lipids: Esters of fatty acids with various alcohols		Complex lipids: Esters of fatty acids containing groups in addition to an alcohol and fatty acids			Precursor and derived lipids
Fats	Waxes	Phospholipids	Glyco-spingolipids	Other complex lipids	
Esters of fatty acids with glycerol, e.g. oils (fats in the liquid form)	Esters of fatty acids with higher molecular weight monohydric alcohols	Contains fatty acids, an alcohol and a phosphoric acid residue and some contain nitrogen containing bases, e.g. glycerophospholipids (alcohol-glycerol) sphingophospholipids (alcohol-sphingosine)	Contains fatty acid, sphingosine and carbohydrates, e.g. galactoceramide	Aminolipids sulfolipids, lipoproteins	Fatty acids, glycerol, other alcohols, steroids, fatty aldehydes, ketones, hydrocarbons, fat-soluble vitamins, hormones

ether, chloroform and benzene into 4 different test tubes. Add one drop of gingili or coconut oil into each tube and shake well.

Observation: Droplet of oil will be seen in the tube containing water. It disappears in other tubes.

Inference: In water, oil broken into droplets and being less dense than water float on the surface. In other solvents, oil dissolves.

Principle: Water is polar in nature whereas oil is hydrophobic. So oil does not dissolve in water but dissolve in fat solvents like ether, chloroform and benzene.

Grease Spot Test

Procedure: Place a drop of gingili oil upon a piece of ordinary writing paper.

Observation: A translucent spot develops.

Inference: Lipids are greasy in nature.

Acrolein Test (Fig. 3.1)

Procedure: Take 2 scoopfuls of potassium bisulfite ($KHSO_4$) in a clean dry test tube. Add 4 drops of gingili oil on the salt and heat gently at first and then more strongly.

Observation: An irritating odor develops.

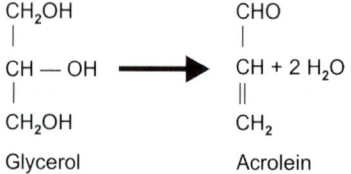

Figure 3.1 Formation of acrolein

Inference: Glycerol part of the oil is dehydrated to acrolein having an irritating odor.

Emulsification Test

Procedure: Add one drop of oil into 2 mL of water in a test tube and shake well. Observe and add 4-6 drops of soap solution and shake well.

Observation: Upon shaking with water, oil floats on the surface. On adding soap solution oil get dispersed in water.

Inference: Surface tension of water is lowered by soap solution and this help the droplets of oil get dispersed and suspended in water (emulsification).

Points to Ponder

Bile salts aid in digestion and absorption of lipids in the gut by emulsification. Bile salts are salts of bile acids and they act as emulsifying agents.

Saponification

Procedure

- Take a clean dry test tube and add 0.5 mL of oil and add 2-3 mL of ethyl alcohol to it and mix well.
- Then add 10 mL of 10% alcoholic NaOH solution. Shake well and keep in a boiling water bath for 15 minutes.
- Take the test tube out of boiling water bath and keep it in a rack for another 15 minutes and add 15 mL of water. Shake thoroughly.
- Divide the contents into 4 equal parts by adding into 4 different tubes marked A, B, C and D.
- To 'A' add 3 mL of concentrated HCl and shake well.
- To 'B' add 4 mL of saturated NaCl solution.
- To 'C' add 3 drops of $CaCl_2$ solution.
- " D " will serve as control.

Observation and Inference (Table 3.2)

Principle: Fats generally composed of esters of fatty acids and can be hydrolyzed to glycerol and fatty acids by different agents like lipase, superheated steam, long continued action of air and light or boiling with alkali. In this test, alkali is used to hydrolyze fat. This process of hydrolyzing triacylglycerol into glycerol and fatty acids by any one of the above said means is known as **saponification.** Metallic salts of higher fatty acids are called **soaps.** Ordinary hard soaps are sodium soaps. Potassium soaps are soft soaps. Calcium and magnesium form insoluble soaps.

Application: The cleansing (detergent) action of soaps is due to their ability to lower surface tension and cause emulsification of oily material, which can then be easily washed away.

Saponification number of a fat is the number of milligrams of KOH required to neutralize free or combined fatty acids in 1 g of fat. It is determined by saponification and titration of excess alkali and is a measure of the mean molecular weight of the fatty acids in a type of fat (e.g. Saponification number of Coconut oil: 253–262; Butter: 210–230).

TABLE 3.2 Observation and inference of saponification

Tube A	White precipitate	Addition of HCl liberates fatty acids which will be seen as white precipitate since it is insoluble in water
Tube B	Pale white layer rises up	Added NaCl reacts with fatty acids to form sodium salts of fatty acids
Tube C	White precipitate	Calcium salt of fatty acid (insoluble in water) are formed on adding $CaCl_2$

$$— CH = CH — + Br_2 \longrightarrow — CHBr — CHBr$$

Figure 3.2 Halogenation

Halogenation Test (Fig. 3.2)

Procedure
- Take two test tubes and mark A and B respectively.
- Add 5 mL of chloroform to both tubes.
  - Add 6–8 drops of oleic acid in tube A and a scoopful of palmitic acid in tube B and shake well.
  - Add a few drops of fresh bromine water in both tubes and shake well.

Observation and Inference (Table 3.3)

Principle: The unsaturated fatty acids possess double bonds and they take up halogens like bromine or iodine at their double bonds.

Application: Iodine number is a measure of amount of unsaturation present in a fat. It is expressed as a number of grams of iodine absorbed by 100 g of fat (e.g. Iodine number of Sunflower oil: 130; Butter: 28).

CHOLESTEROL

Identification by microscopy: Cholesterol has a characteristic shape which can be seen by microscopy. Cholesterol crystals have a rhombic shape with a notch at one corner (Fig. 3.3).

TABLE 3.3 Observation and inference of Halogenation test

Tube A	Orange yellow color of bromine water vanishes	Oleic acid acid (C 18:1) is an unsaturated fatty acid and take up bromine atoms at the double bonds
Tube B	Orange yellow color	Palmitic acid is a saturated fatty acid (C 16), hence cannot take up bromine atoms due to the absence of double bonds

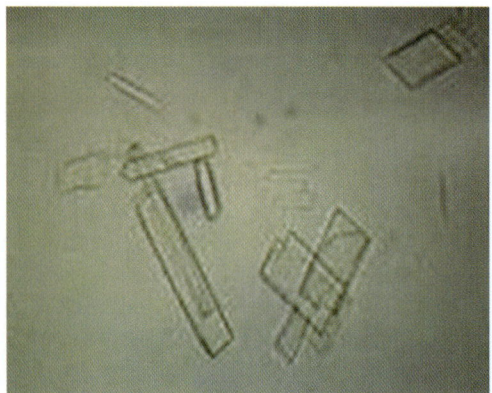

Figure 3.3 Microscopy of cholesterol crystals

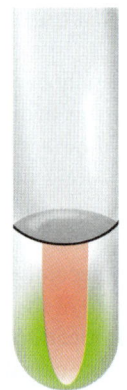

Figure 3.4 Salkowski's reaction

Application: Microscopic examination of body fluids suspected to contain cholesterol can be detected by microscopy by the characteristic rhombic crystals, notched at one corner.

Color Reactions of Cholesterol

Salkowski's Reaction (H_2SO_4 Test) (Fig. 3.4)

Procedure: Dissolve a few crystals of cholesterol in 2 mL of chloroform in a dry test tube and add an equal volume of concentrated H_2SO_4 gently along the sides of the tube. The acid being heavier goes down.

Observation: A play of colors from bluish red to cherry red to purple and finally sulfuric acid becomes red with green fluorescence.

Inference: Cholesterol is dehydrated by concentrated sulfuric acid to form 3, 5, cholestadiene or 2, 4 cholestadiene. They polymerize and react with sulfuric acid to form their sulfuric acid derivatives giving rise to a play of colors.

Libermann-Burchard Reaction (Acetic Anhydride Sulfuric Acid Test) (Fig. 3.5)

Procedure: Dissolve a few crystals of cholesterol in 2 mL of chloroform in a dry test tube. Add 10 drops of acetic anhydride and 1–3 drops of concentrated sulfuric acid.

Observation: The solution becomes red, then blue and finally bluish green in color.

Principle: Blue green product of Libermann-Burehard (LB) reaction is an oxidation product of cholesterol → 3, 5 cholestadiene.

Points to ponder

Dry glassware must be used for cholesterol experimentation.

Figure 3.5 Libermann-Burchard reaction (LB reaction)

(Bluish green color)

QUESTIONS

1. Name the following:
   a. Derived lipids
   b. Complex lipids
   c. Phospholipids
   d. Emulsifying agent of digestion and absorption of lipids in the intestine
   e. Test to detect the presence of glycerol in fat
   f. Insoluble soap
   g. Soft soap
   h. Test to detect the presence of unsaturated fatty acid
   i. Two reactions to detect cholesterol
2. Give the principle of the following tests:
   a. Acrolein test
   b. Emulsification test
   c. Libermann-Burchard reaction
   d. Salkowski's reaction
   e. Halogenation test
3. Give brief answers:
   a. Classify lipids.
   b. What are oils.
   c. Role of emulsification of fat in digestion.
   d. What is saponification and saponification number?
   e. Iodine number.

REAGENT PREPARATION

- **Chloroform:** As such from the container, used for laboratory purpose.
- **Ether:** As such from the container, used for laboratory purpose.
- **Benzene:** As such from the container used for laboratory purpose.
- **Oil:** Gingili, sunflower oil or coconut oil.
- **Potassium bisulfite:** Solid potassium bisulfite as such is used.
- **Ethyl alcohol:** As such from the container used for laboratory purpose.
- **Saturated NaCl:** Take water in a beaker or cylinder. Dissolve sodium chloride until some crystals remain undissolved.
- **Bromine water:** Add a few drops of liquid bromine (caution: very corrosive) to 100 mL water. Prepare fresh (Fading of color indicates, inactive reagent).
- **Palmitic acid:** As such from the container used for laboratory purpose.
- **Oleic acid:** As such from the container used for laboratory purpose.
- **Acetic anhydride:** As such from the container used for laboratory purpose.
- **Concentrated HCl:** As such from the container used for laboratory purpose.

CHAPTER 4

Reactions of Urea

GENERAL REACTIONS OF UREA

Introduction

Urea is the end product of protein catabolism (Fig. 4.1). It is a non-protein nitrogen (NPN). Non-protein nitrogen of the blood and other body fluids include constituents, which are not precipitated as proteins, e.g. Major NPNs of blood urea, uric acid, creatinine, creatine, amino acids and glutathione.

The sources of the constituent atoms of urea (H_2N-CO-NH_2)—carbon from carbon dioxide, one nitrogen from ammonia and the other nitrogen from aspartate. The full set of enzymes required for the formation of urea present only in hepatocytes (liver cells). Hence, urea is synthesized exclusively in the liver. It is mainly excreted in urine, i.e. 80–90% of total nitrogen of human urine is present as urea. *Unlike other NPNs, the concentration of urea in urine decreases* when protein intake is restricted. The amount of urea excreted per day by a normal adult is about 15–30 g per day. Urea content is increased with high protein diet. Blood urea level decreases in severe liver diseases since it is the sole organ concerned with its synthesis.

Physical Properties of Urea Solution

- Appearance   clear
- Color   colorless
- Odor   odorless
- Reaction to litmus No change

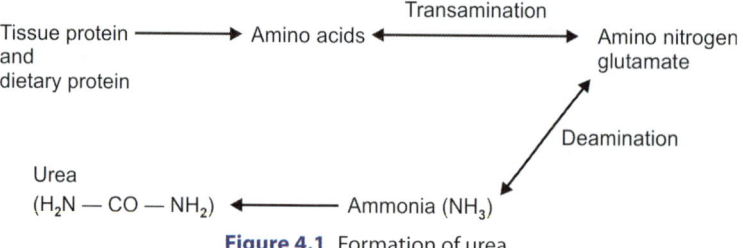

Figure 4.1 Formation of urea

Chemical Reactions

Alkaline Hypobromite Test

Procedure: To 3 mL of urea solution add a few drops of alkaline hypobromite solution (3 mL concentrated NaOH + 2 mL bromine water).

Observation: Brisk effervescence.

Inference: Nitrogen present in urea is liberated as nitrogen gas.

Principle: Sodium hypobromite decomposes urea to CO_2 and N_2. Carbon dioxide is absorbed by the excess sodium hydroxide and the nitrogen is evolved and causing the brisk effervescence. (Fig. 4.2).

Points to Ponder: All ammonium compounds and all compounds containing amino group ($-NH_2$) release N_2 when treated with alkaline hypobromite.

Specific Urease Test (Figs 4.3 and 4.4)

Procedure
- To 2 mL urea solution in a test tube add a drop of phenol red indicator.
- Add 2% Na_2CO_3 solution drop by drop till a pink color develops.
  (pH range of Phenol red: 6.8–8.4 and color range: yellow to red)
- Add 2% acetic acid drop by drop till the pink color just disappears indicating the pH nearer to 6.8.
- Add a 1 mL of urease enzyme extract.
- Keep at 37°C in an incubator for 5 minutes because optimum temperature of urease enzyme is 37°C.
- Set a control tube taking the entire ingredients but using boiled urease enzyme extract.

Observation: Pink color develops in the tube containing urea and no change in control tube.

Inference: Urease decomposes urea to ammonium carbonate. Ammonium carbonate being basic raises the pH. Phenol red used in this test will show pink to red color at basic pH.

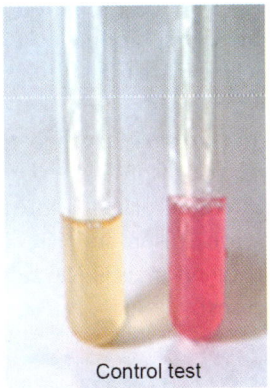

Figure 4.3 Specific urease test

$$\text{Urea} \xrightarrow[\text{Urease}]{H_2O} \text{Ammonium carbonate}$$

Figure 4.4 Urease action on urea

$$H_2N-CO-NH_2 + 3\, NaBrO$$
$$\downarrow$$
$$NaBr + N_2 + CO_2 + 2H_2O$$

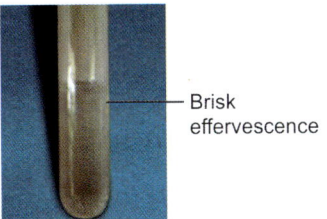
Brisk effervescence

Figure 4.2 Alkaline hypobromite test and reaction

QUESTIONS

1. Answer to the point:
   a. Number of nitrogen atoms in urea
   b. Sources of nitrogen atoms in urea
   c. Source of carbon in urea
   d. Cells producing urea in the body
   e. Concentration of urea in blood in a normal adult
   f. Two tests to detect urea from a specimen
   g. Main route of excretion of urea from the body
   h. Three non-protein nitrogen compounds in the blood
2. Give the principle of the following tests:
   a. Alkaline hypobromite test
   b. Specific urease test
3. Give brief answers:
   a. Formation of urea by hepatocytes
   b. Urea cycle disorders
   c. Biochemical principle of treatment of urea cycle disorders
   d. Why blood urea is elevated in kidney diseases?

REAGENT PREPARATION

- **Sodium hydroxide concentrated (1N):** Dissolve 40 g reagent grade sodium hydroxide pellets in a few mL of distilled water to make 1 liter. Cool. Allow to stand for 3 days or so. Decant the solution into a bottle fitted with siphon and a calcium chloride tube to prevent entry of carbon dioxide. Standardize by titration with an acid of known strength using methyl red as indicator (secondary standardization) or potassium biphthalate of known strength (primary standardization).
- **Sodium hypobromite (Alkaline hypobromite):** Mix 25 mL of liquid bromine with 250 mL of 40% NaOH.
- **2% acetic acid:** Dissolve 2 mL of glacial acetic acid (99.8%) in 100 mL of water.
- **2% Na_2CO_3**
- **Urease enzyme:** Grind 10 g horse gram or jack bean or soybean with 100 mL 30% alcohol and take the extract.
- **Phenol red indicator:** Dissolve 1 g phenol red in 100 mL distilled water. Add 5.7 mL of 0.05 N NaOH and make the volume up to 250 mL with water.

CHAPTER

5

Reactions of Creatinine

GENERAL REACTIONS OF CREATININE

Introduction

Creatinine is an anhydride of creatine (Fig. 5.1). It is a constituent of normal human urine. It is a non-protein nitrogen (NPN). Non-protein nitrogen of the blood and other body fluids include the constituents which are not precipitated as proteins, e.g. Some NPN of blood are urea, uric acid, creatinine, creatine, amino acids, glutathione and other compounds.

Blood level of creatinine is 0.7–1.2 mg%. Rate of excretion of creatinine in urine in an adult is 1–1.8 g/day. Foods such as meat and fish contain significant amount of creatinine especially after cooking. Hence, fasting blood sample is ideal for creatinine estimation (fasting creatinine). The blood level of creatinine on a creatinine free diet is almost constant for a given individual and is independent of the total nitrogen excreted in these conditions. It can be expressed as creatinine coefficient or the daily excretion of creatinine in mg per kg body weight. The endogenous creatinine is formed from creatine phosphate which is concerned with muscle contraction. Creatinine is the least variable non-protein nitrogenous (NPN) constituent of the blood.

Physical Properties

Appearance	-	clear
Color	-	colorless
Odor	-	odorless
Reaction to litmus	-	acidic (the creatinine solution provided in mild acid medium)
Solubility	-	slightly soluble of cold water, more soluble in warm water and in warm alcohol

Reactivity: It forms salts with strong mineral acids.

Chemical Properties

Nitroprusside Test

Procedure: Take 5 mL of creatinine solution in a test tube add a few drops of sodium nitroprusside and render the solution alkaline by adding dilute NaOH. A ruby red color appears

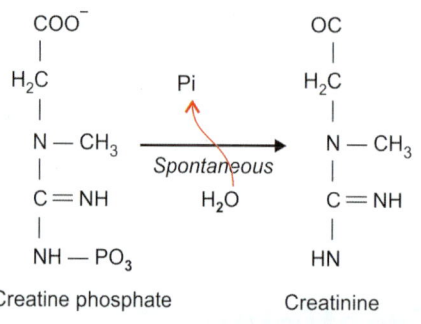

Figure 5.1 Formation of creatinine from creatine phosphate

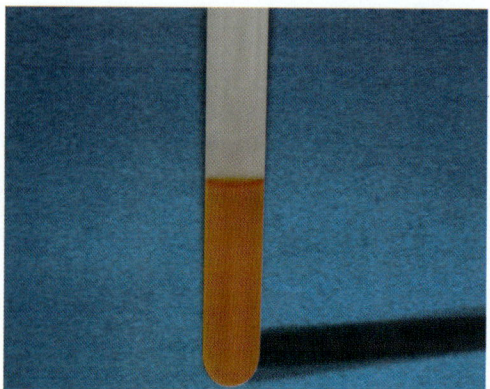

Figure 5.2 Jaffe's test

and soon turns yellow. To this yellow solution add an excess of acetic acid and apply heat.

Observation: A green color forms and changes to blue color.

Inference: The color is due to the formation of Prussian blue.

Jaffe's Test (Picric Acid Reaction)

Procedure: To 5 mL of creatinine solution add 1 mL of 1% picric acid and 10 drops of 10% NaOH. Shake well and keep it for a few minutes.

Observation: An orange red color forms (Fig. 5.2).

Inference: Creatinine forms creatinine picrate in alkaline medium which is red in color.

QUESTIONS

1. Give an outline of formation of creatinine.
2. What is the reference range of creatinine in the blood?
3. Give the rate of excretion of creatinine in urine.
4. On creatinine free diet it is excretion in urine is constant in an adult individual. Explain.
5. What are the creatinine-rich food items?
6. What is creatinine clearance test? What is its significance in clinical diagnosis.
7. What is the reaction used to estimate the creatinine in body fluids?
8. Give the application of Jaffe's test.

REAGENT PREPARATION

- **Creatinine solution (0.1%):** Dissolve 100 mg (0.1 g) creatinine in 0.1 N HCl and make up to 100 mL.
- **Picric acid (1%):** Dissolve 1 g anhydrous picric acid in 100 mL distilled water.

CHAPTER 6

Reactions of Uric Acid

GENERAL REACTIONS OF URIC ACID

Introduction

Uric acid is filtered by the glomeruli. Almost all of the uric acid filtered is reabsorbed at proximal convoluted. Uric acid (Figs 6.1 and 6.2) is the end product of purine catabolism. It is a non-protein nitrogen (NPN). Non-protein nitrogen of the blood and other body fluids include the constituents which are not precipitated as proteins, e.g. Some NPN of blood are urea, uric acid, creatinine, creatine, amino acids, glutathione.

Blood level of uric acid 2–7 mg%. Over production of uric acid in the body is due to increased formation and break down of purine nucleotides. Uric acid in the plasma is filtered by the glomeruli. Almost all the uric acid filtered, is reabsorbed at the proximal convoluted tubules (PCT) and some secreted at the distal part of the PCT and there is further reabsorption of uric acid in distal convoluted tubules (DCT). Final urine contains about 10% of the filtered uric acid. Above pH 5.75 most of the uric acid molecules are ionized as urate ion and are more soluble than nonionized uric acid seen at pH below 5.75. This kind of solubility exists in urine as well as in other body fluids.

Rate of excretion of uric acid in an adult is about 0.5–1 g in urine per day. But it varies with purine content of the diet. The uric acid content of the urine is relevant in relation to the formation of uric acid calculi. Intake of alkali, carbonates and citrates or base forming foods increase the pH of urine and enhance the solubility of uric acid in urine and prevent calculi formation.

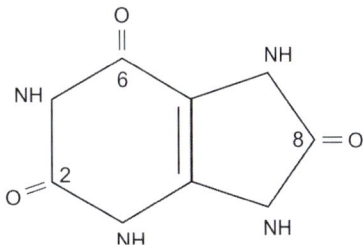

Figure 6.1 Uric acid (2,6,8 Tri oxy purine)

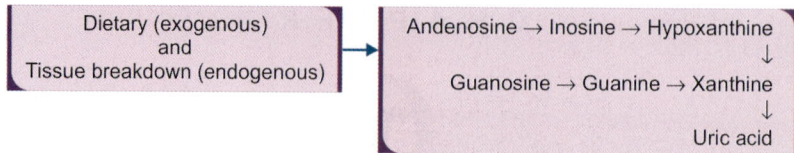

Figure 6.2 Formation of uric acid from purine nucleotides

Physical Properties

Appearance	Transparent
Color	White
Odor	Odorless
Reaction to litmus	Alkaline (uric acid dissolved in mild alkaline medium)
Solubility	Insoluble in alcohol, ether; soluble in boiling water to some extent, insoluble in cold water; soluble in alkali and concentrated sulfuric acid

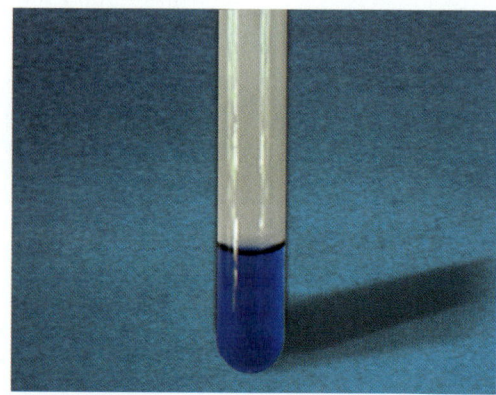

Figure 6.3 Benedict's uric acid test

Chemical Reactions

Benedict's Uric Acid Test (Fig. 6.3)

Procedure: To 5 mL of test solution add 1 mL of 1% Na_2CO_3 and a few drops of Benedict's uric acid reagent.

Observation: Intense blue color.

Inference: Uric acid reduces phosphotungstic acid to tungsten blue in alkaline medium.

Schiff's Test

Procedure: Add a drop of 3% ammoniacal silver nitrate at the center of a circular filter paper. Then add a drop of uric acid solution on to it.

Observation: Black color appears (Fig. 6.4).

Inference: Uric acid reduces silver nitrate to metallic silver in alkaline medium.

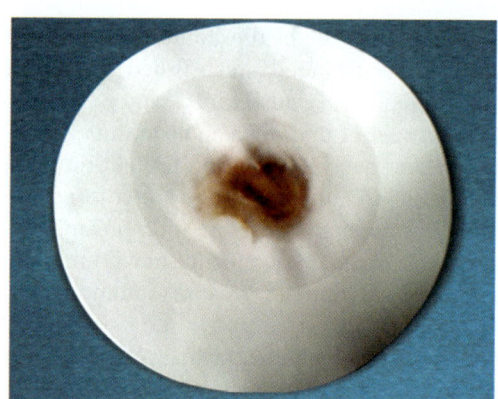

Figure 6.4 Schiff's test

Murexide Test (Fig. 6.5)

Procedure: Add 2 or 3 drops of concentrated nitric acid to a pinch of uric acid in a small evaporating dish and evaporate to dryness by heating on a water bath. A red or yellow residue forms. Then add a drop of a dilute ammonium hydroxide to one edge of the residue and to

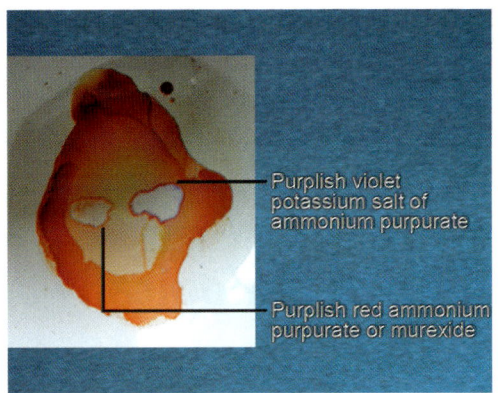

Figure 6.5 Murexide test

the opposite edge, add a drop of potassium hydroxide.

Observation: Purplish red color with ammonium hydroxide and purplish violet color with potassium hydroxide.

Principle: In this reaction, uric acid is oxidized to dialuric acid and alloxan which condense to form alloxantin. The alloxantin so formed reacts with ammonium hydroxide to form ammonium purpurate or murexide which is purplish red in color. With potassium hydroxide a purplish violet color is produced due to the formation of potassium salt of ammonium purpurate.

Application: Murexide test is useful to detect uric acid calculi.

QUESTIONS

1. Give an outline of formation of uric acid in human body.
2. What is the reference range of uric acid in the blood?
3. Give the rate of excretion of uric acid in urine.
4. Uric acid tend to form stone easily in acid pH of urine. Explain.
5. Give the principle of Benedict's uric acid test.
6. Give the principle of Schiff's test.
7. What is the chemistry of murexide test?
8. Application of murexide test.

REAGENT PREPARATION

- **Benedict's uric acid reagent:** Dissolve 100 g of pure sodium tungstate in a few mL water in a one liter pyrex glass flask and add 50 g pure arsenic acid (arsenic pentoxide—As_2O_5) followed by 25 mL 85% phosphoric acid and 20 mL of concentrated HCl, boil for 20 minutes, cool and make up to 1 L.
- **Na_2CO_3 1%:** Dissolve 1 g sodium carbonate in 100 mL of water.
- **Ammoniacal silver nitrate 3%:** Dissolve 26 g of silver nitrate in about 500 mL water, add enough ammonium hydroxide to redissolve the precipitate which forms upon the addition of ammonium hydroxide initially. Then make up the volume to 1 L with water.

CHAPTER 7

Schemes for Identification of Biologically Important Compounds

FOR BIOLOGICALLY IMPORTANT CARBOHYDRATE SUBSTANCES COMPRISING GLUCOSE, FRUCTOSE, SUCROSE, LACTOSE, MALTOSE

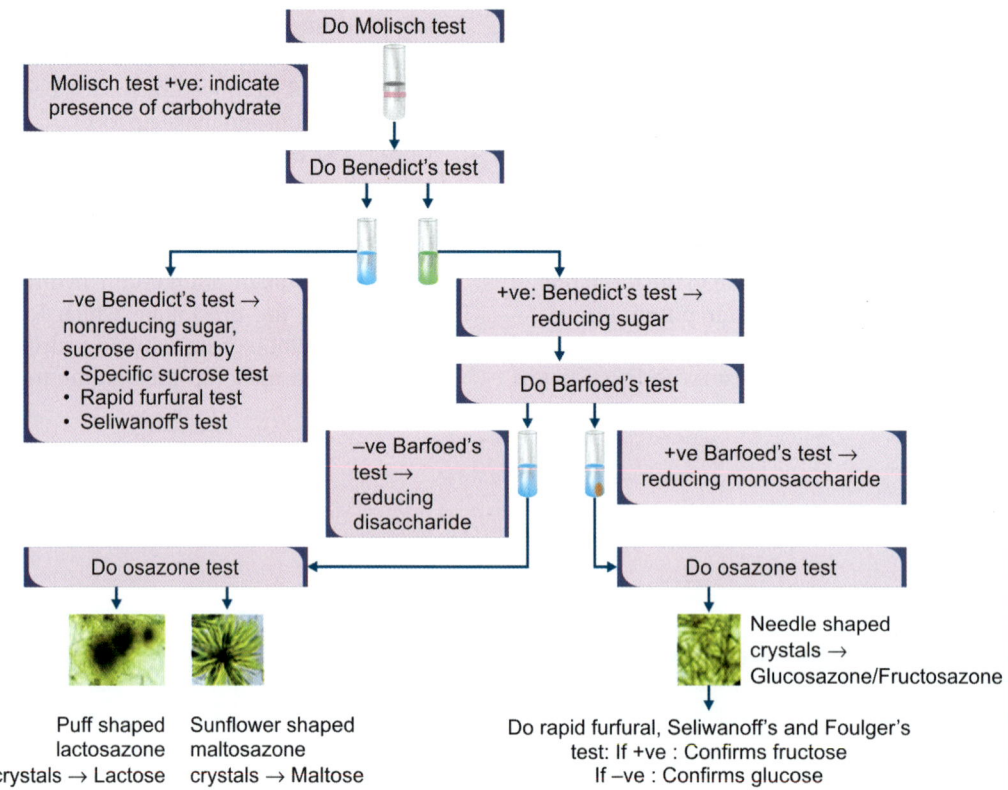

Chapter 7 Schemes for Identification of Biologically Important Compounds

FOR BIOLOGICALLY IMPORTANT SUBSTANCES COMPRISING GLUCOSE, FRUCTOSE, SUCROSE, LACTOSE, MALTOSE, ALBUMIN, UREA, URIC ACID AND CREATININE

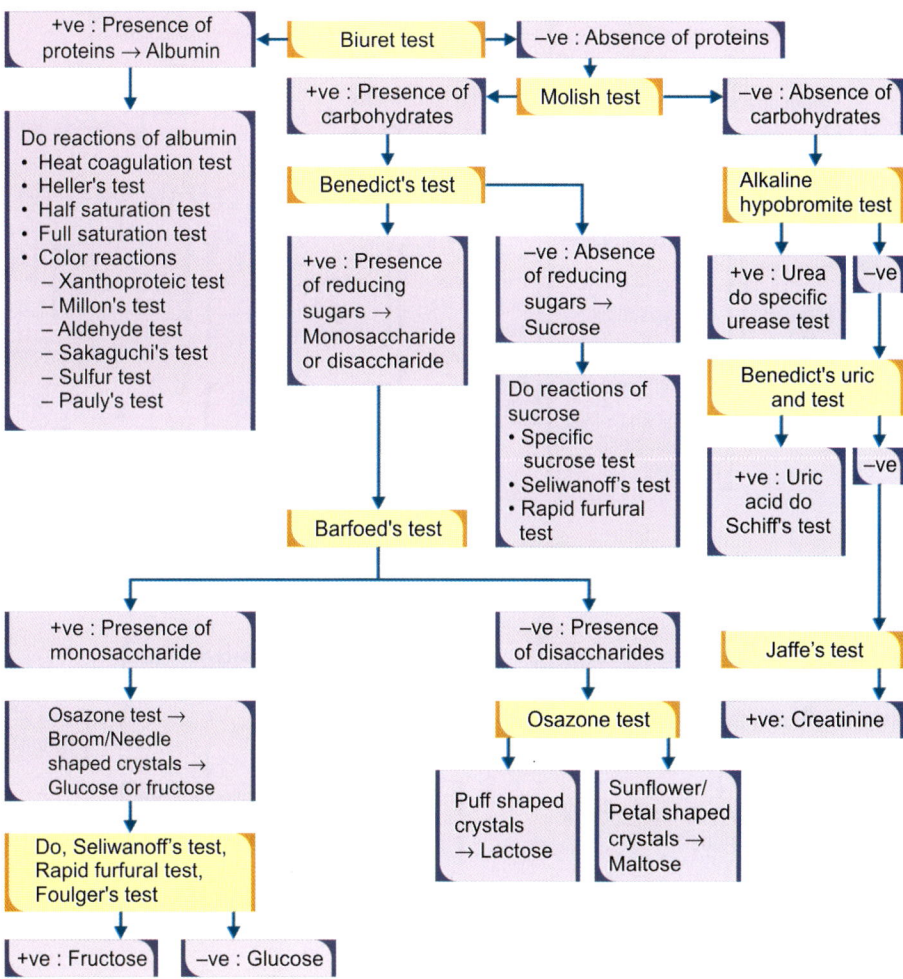

FOR BIOLOGICALLY IMPORTANT SUBSTANCES COMPRISING GLUCOSE, FRUCTOSE, SUCROSE, LACTOSE, MALTOSE, ALBUMIN, CASEIN, UREA, URIC ACID AND CREATININE

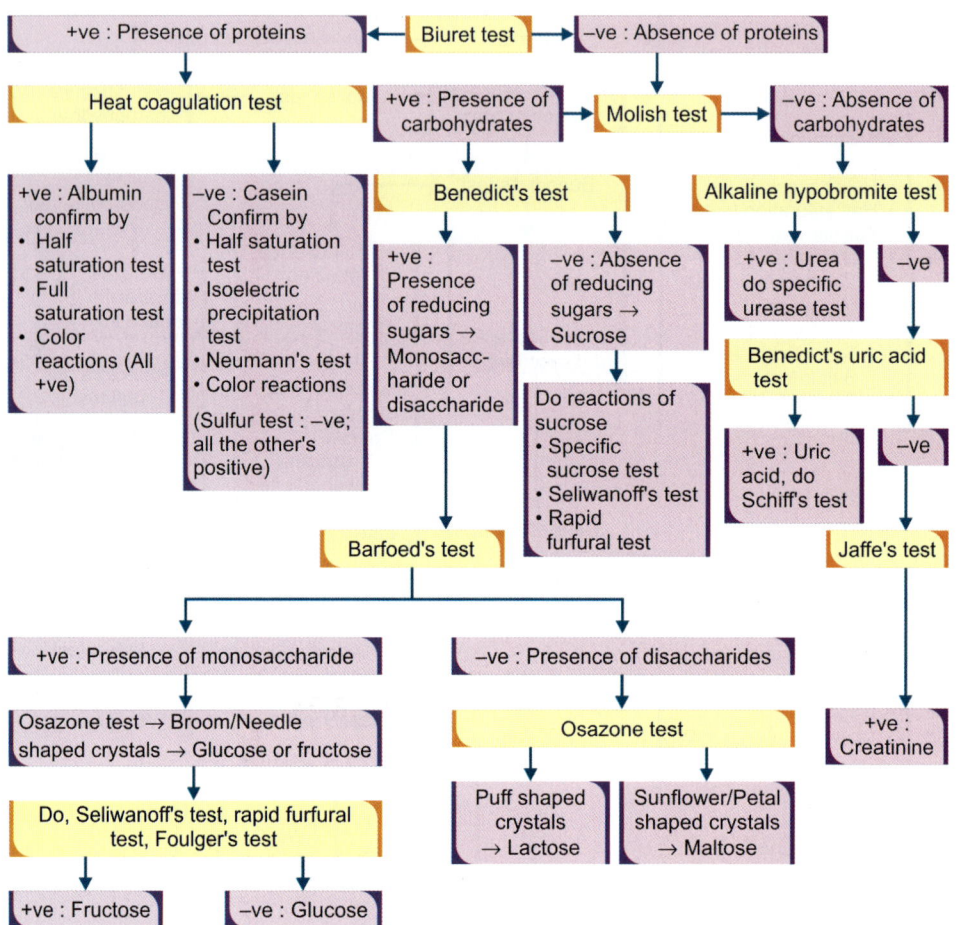

CHAPTER 8

Urine Analysis

ANALYSIS OF NORMAL CONSTITUENTS OF URINE

Introduction

Urine is the ultra-filtrate of plasma formed when the blood perfuse the two kidneys. Glomerulus filters plasma and the volume of glomerular filtrate amount to 180 L in 24 hours for an adult. Tubules of the kidney modify the glomerular filtrate by reabsorption and secretion of water and solutes to produce final urine volume of 1–2 L per day. Glomerular filtration rate is about 120 mL per minute. Thus, the kidneys retain essential substances and excrete waste products from the body. By this process, it also helps in maintaining the acid base balance.

Clinical laboratory analysis of urine can provide information of kidney dysfunction (e.g. nephrotic syndrome, glomerulonephritis) and about several metabolic diseases (e.g. phenyl ketonuria, diabetes mellitus) in an individual.

Specimen Collection

For getting correct analytical results, care must be taken in the collection of urine and transportation of it to the laboratory. Urine should be collected in clean sterile containers and labeled with name, age, date and time of collection.

The best urine specimen is the first voided urine (midstream specimen) in the morning since it is the most concentrated.

Physical Examination of Urine (Table 8.2)

Physical examination of urine is to be carried out prior to routine analysis which includes assessment of **volume, appearance, odor, color, pH and specific gravity.** Careful interpretation of these physical properties gives us a lot of information regarding various types of illnesses.

Volume

Normal adults excrete about 750–2000 mL of urine. It is influenced by fluid and salt intake, perspiration, respiration and functional status of cardiovascular and renal systems.

Oliguria

A decreased urine output (< 400 mL/day) is called oliguria.

TABLE 8.1 Diseases associated with defective renal concentrating mechanism → low specific gravity

Marked polyuria and hypotonic urine after water deprivation*	Moderate polyuria and inability to produce hypertonic urine
Diabetes insipidus (Hypothalamic or pituitary disorders → ADH deficiency)	Hypercalcemia
	Hypokalemia
	Chronic pyelonephritis
Chronic lithium toxicity	End stage kidney disease
Sickle cell nephropathy	Amyloidosis
	Interstitial nephritis

*Water deprivation for 18 hours[1]

Causes of oliguria:
- Prerenal causes → low blood pressure, shock, bleeding, fluid deprivation.
- Renal causes → acute tubular necrosis, poisons causing renal damage, renal vascular disease.
- Postrenal causes → calculi, tumors compressing urinary tract from within or outside, prostate enlargement.

Polyuria (Table 8.1)

An increased output (>3000 mL/day) of urine is referred to as polyuria.

Causes of polyuria:
- Conditions leading to excretion of a large amount of solutes along with iso-osmotic amount of water, e.g. excessive salt intake, diabetes mellitus
- Deficiency of antidiuretic hormone (ADH)
- Excessive fluid intake
- Intake of diuretics.

Appearance

Normal urine is clear (transparent).

Causes of Cloudiness
- Presence of amorphous **phosphates** or **urates** in urine. The cloudiness caused by phosphate appears on heating and that due to urates disappear upon heating.
- Pus cell (white blood cells) clears on filtering.
- Bacteria or fungi cleared by centrifugation.
- Colloidal suspension of fat (as in chyluria) which cannot be cleared off by usual filtering or centrifugation.

Odor

Normally fresh urine has a faint aromatic smell.
- Upon standing strong ammoniacal odor develop due to formation of ammonia by the decomposition of urea.
- Presence of ketone bodies like acetone in urine produces a fruity odor.

Color

Normal color of urine varies from colorless to deep yellow. The color of urine is conferred by urochromes and urobilin. The intensity of the color varies with degree of dilution—dilute urine is pale yellow and concentrated urine is deep yellow.

Change in color of urine is observed in different clinical conditions—a few examples are given below:
- Deep or brownish yellow → bile pigments (jaundice)
- Red color → intact red cells, free hemoglobin or myoglobin
- Black color → alkaptonuria, melanuria.

Chapter 7 Schemes for Identification of Biologically Important Compounds

FOR BIOLOGICALLY IMPORTANT SUBSTANCES COMPRISING GLUCOSE, FRUCTOSE, SUCROSE, LACTOSE, MALTOSE, ALBUMIN, UREA, URIC ACID AND CREATININE

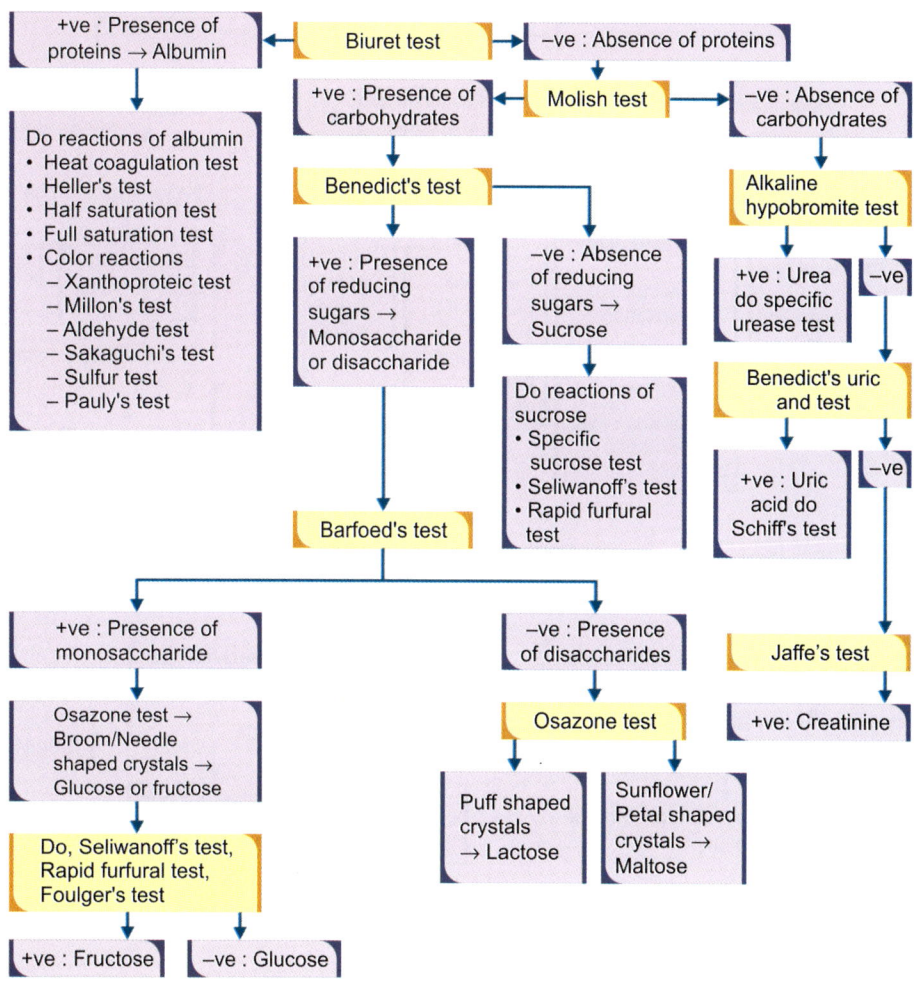

FOR BIOLOGICALLY IMPORTANT SUBSTANCES COMPRISING GLUCOSE, FRUCTOSE, SUCROSE, LACTOSE, MALTOSE, ALBUMIN, CASEIN, UREA, URIC ACID AND CREATININE

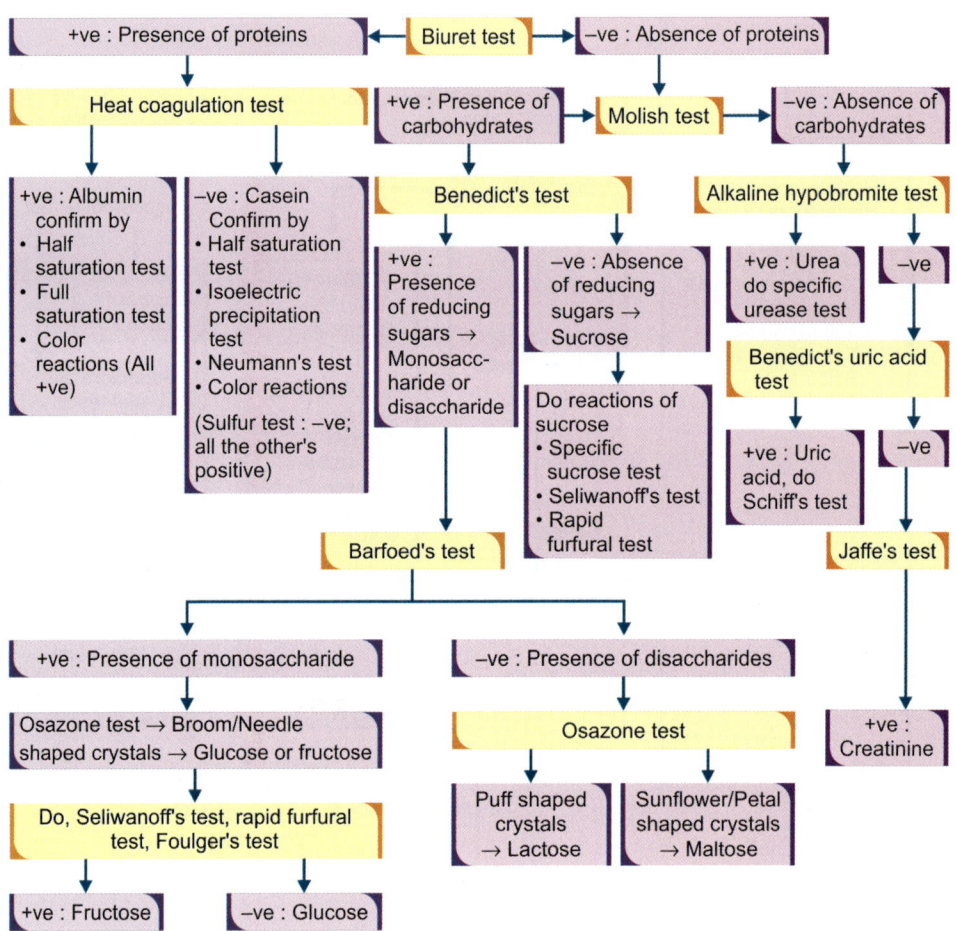

CHAPTER 8

Urine Analysis

ANALYSIS OF NORMAL CONSTITUENTS OF URINE

Introduction

Urine is the ultra-filtrate of plasma formed when the blood perfuse the two kidneys. Glomerulus filters plasma and the volume of glomerular filtrate amount to 180 L in 24 hours for an adult. Tubules of the kidney modify the glomerular filtrate by reabsorption and secretion of water and solutes to produce final urine volume of 1-2 L per day. Glomerular filtration rate is about 120 mL per minute. Thus, the kidneys retain essential substances and excrete waste products from the body. By this process, it also helps in maintaining the acid base balance.

Clinical laboratory analysis of urine can provide information of kidney dysfunction (e.g. nephrotic syndrome, glomerulonephritis) and about several metabolic diseases (e.g. phenyl ketonuria, diabetes mellitus) in an individual.

Specimen Collection

For getting correct analytical results, care must be taken in the collection of urine and transportation of it to the laboratory. Urine should be collected in clean sterile containers and labeled with name, age, date and time of collection.

The best urine specimen is the first voided urine (midstream specimen) in the morning since it is the most concentrated.

Physical Examination of Urine (Table 8.2)

Physical examination of urine is to be carried out prior to routine analysis which includes assessment of **volume, appearance, odor, color, pH and specific gravity.** Careful interpretation of these physical properties gives us a lot of information regarding various types of illnesses.

Volume

Normal adults excrete about 750-2000 mL of urine. It is influenced by fluid and salt intake, perspiration, respiration and functional status of cardiovascular and renal systems.

Oliguria
A decreased urine output (< 400 mL/day) is called oliguria.

TABLE 8.1 Diseases associated with defective renal concentrating mechanism → low specific gravity

Marked polyuria and hypotonic urine after water deprivation*	Moderate polyuria and inability to produce hypertonic urine
Diabetes insipidus (Hypothalamic or pituitary disorders → ADH deficiency)	Hypercalcemia
	Hypokalemia
	Chronic pyelonephritis
Chronic lithium toxicity	End stage kidney disease
Sickle cell nephropathy	Amyloidosis
	Interstitial nephritis

*Water deprivation for 18 hours[1]

Causes of oliguria:
- Prerenal causes → low blood pressure, shock, bleeding, fluid deprivation.
- Renal causes → acute tubular necrosis, poisons causing renal damage, renal vascular disease.
- Postrenal causes → calculi, tumors compressing urinary tract from within or outside, prostate enlargement.

Polyuria (Table 8.1)

An increased output (>3000 mL/day) of urine is referred to as polyuria.

Causes of polyuria:
- Conditions leading to excretion of a large amount of solutes along with iso-osmotic amount of water, e.g. excessive salt intake, diabetes mellitus
- Deficiency of antidiuretic hormone (ADH)
- Excessive fluid intake
- Intake of diuretics.

Appearance

Normal urine is clear (transparent).

Causes of Cloudiness
- Presence of amorphous **phosphates** or **urates** in urine. The cloudiness caused by phosphate appears on heating and that due to urates disappear upon heating.
- Pus cell (white blood cells) clears on filtering.
- Bacteria or fungi cleared by centrifugation.
- Colloidal suspension of fat (as in chyluria) which cannot be cleared off by usual filtering or centrifugation.

Odor

Normally fresh urine has a faint aromatic smell.
- Upon standing strong ammoniacal odor develop due to formation of ammonia by the decomposition of urea.
- Presence of ketone bodies like acetone in urine produces a fruity odor.

Color

Normal color of urine varies from colorless to deep yellow. The color of urine is conferred by urochromes and urobilin. The intensity of the color varies with degree of dilution—dilute urine is pale yellow and concentrated urine is deep yellow.

Change in color of urine is observed in different clinical conditions—a few examples are given below:
- Deep or brownish yellow → bile pigments (jaundice)
- Red color → intact red cells, free hemoglobin or myoglobin
- Black color → alkaptonuria, melanuria.

pH

In a healthy person, the pH of the urine varies from 4.6 to 8 depending on many factors like dietary intake and metabolic activities. Most often the urine pH is acidic around 6.0 due to the presence of sulfates, phosphates, chlorides and nonvolatile organic acids.

Vegetarian diet produces alkaline urine. On keeping urine, it becomes alkaline due to the formation of ammonia formed by the decomposition of urea.

Measurement of pH
- *Litmus paper:* In acid urine, blue litmus turns red and in alkaline urine, red litmus turns blue.
- *pH paper* which has a wide range of colors from 4.5 to 7.5.
- *Dip sticks* uses a combination of indicators methyl red and bromophenol blue which give a range of different colors from orange to green to blue as the pH rises from pH 5.0 to pH 9.0.

Points to Ponder: Extremely acidic or alkaline urine suggest the possibility of poorly collected urine.

Specific Gravity

Specific gravity of urine serves to assess the **concentrating ability** of the kidneys. Under normal conditions the specific gravity of urine range from 1.015 to 1.025. Normally specific gravity of urine varies widely depending on hydration status. Excessive fluid intake → fall in specific gravity and severe dehydration → rise in specific gravity.

If urine contains abnormal constituents like proteins and glucose, specific gravity will be increased and a correction factor should be applied to compensate for these factors.

Correction factor for proteins: Subtract 0.003 from the specific gravity reading, for each 1 g of protein/dL of urine.

Correction factor for glucose: Subtract 0.004 from the specific gravity reading for each 1 g of glucose/dL of urine.

Fixed specific gravity: The specific gravity of urine is identical to the glomerular filtrate around 1.010. It is seen in patients with chronic kidney disease (CKD).

Presence of substances with high molecular weight substances like proteins and glucose in the urine impart much higher specific gravity than due to the excessive excretion of crystalloids.

Measurement of specific gravity is done by **urinometer** (see it in the section - spotters).

Better method for determining concentrating ability of kidneys: Measurement of osmotic concentration of urine is better because regulation of water excretion depends partly on osmolality of the fluid compartments of the body. **Refractometer** used for measuring urine osmolality.

Demonstration of Inorganic Constituents of Urine (Table 8.3)

The main inorganic constituents of urine are Na^+, K^+, Ca^{++}, Mg^{++}, NH_4^+, Cl^-, phosphates and sulfates.

- Chief inorganic constituent of urine is **chloride**. It is derived from salts of the diet. Rate of excretion of chloride in urine is **10–15 g per day**. Its content in urine is increased in Addison's disease in which there is

TABLE 8.2 Physical properties of normal urine

S. no.	Experiment	Observation
1.	Appearance	Clear
2.	Color	Amber yellow
3.	Odor	Ammoniacal smell
4.	Reaction to litmus	Blue litmus turns red (mostly acidic)
5.	Specific gravity	1.015–1.025

aldosterone deficiency so that reabsorption sodium and chloride are defective → increased excretion in urine.
- The **sulfates** of urine derived from sulfur containing amino acids. Rate of excretion of inorganic sulfates in urine **0.8–1g/day**.
- **Calcium** is excreted at the rate of **0.1–0.3 g/day**.
- **Phosphates** derived from inorganic phosphates in the diet—phosphoproteins, nucleoproteins and phospholipids. It is excreted at the rate of **1g per day.**

Test for Chloride

Procedure: Acidify 2 mL of urine with 2 drops of concentrated HNO_3 and add 2 mL of silver nitrate solution.

Observation: White precipitate.

Inference: A white precipitate of silver chloride (AgCl) forms. Nitric acid prevents precipitation of salts other than chloride like silver urates and silver phosphates.

Test for Sulfates

Procedure: Acidify 3 mL of urine with 2–4 drops of concentrated HCl and add 1 mL of barium chloride solution.

Observation: White precipitate.

Inference: A white precipitate of barium sulfate. HCl prevents precipitation of phosphates.

Test for Calcium and Phosphates (Fig. 8.1)

Procedure:
- Take 10–12 mL of urine in a boiling tube add 3 mL of strong ammonia solution and boil till white precipitates of calcium and magnesium are formed.
- Filter through a filter paper placed in a funnel placed over a test tube.
- Wash the precipitate thus collected in the filter paper by just pouring a few mL of water through the filter paper.
- Then transfer the funnel with the filter paper to another test tube.
- Add 3 mL hot acetic acid through the filter paper placed over the test tube to dissolve the precipitate in the filter paper and to collect it in the test tube underneath.
- Divide it into 2 parts:
  1. *To detect calcium:* To one part add 1 mL of potassium oxalate.
  2. *To detect phosphates:* To the other part add a drop of concentrated HNO_3 and a few drops of ammonium molybdate solution. Boil.

Observation: White precipitate forms in the test meant for **calcium.** Fine lemon yellow (canary yellow) precipitate forms in the tube meant for detecting **phosphates.**

Inference: Calcium forms a white precipitate of calcium oxalate on addition of potassium oxalate. On boiling with ammonium hydroxide, phosphates of calcium and magnesium are precipitated. These are then filtered and redissolved in hot acetic acid. Phosphates react with ammonium molybdate to form canary yellow colored **ammonium phosphomolybdate** in the presence of HNO_3.

Test for Ammonia

Procedure: To 10 mL of urine add a drop of phenolphthalein and make just alkaline by adding 0.1 N NaOH in drops. Hold a glass rod dipped in phenolphthalein at the mouth of the test tube and heat the contents of the tube.

Observation: The phenolphthalein indicator at the tip of the glass rod turns pink.

Inference: Ammonium salts release ammonia in alkaline medium and the ammonia vapors emerging from the tube turn the phenolphtha-

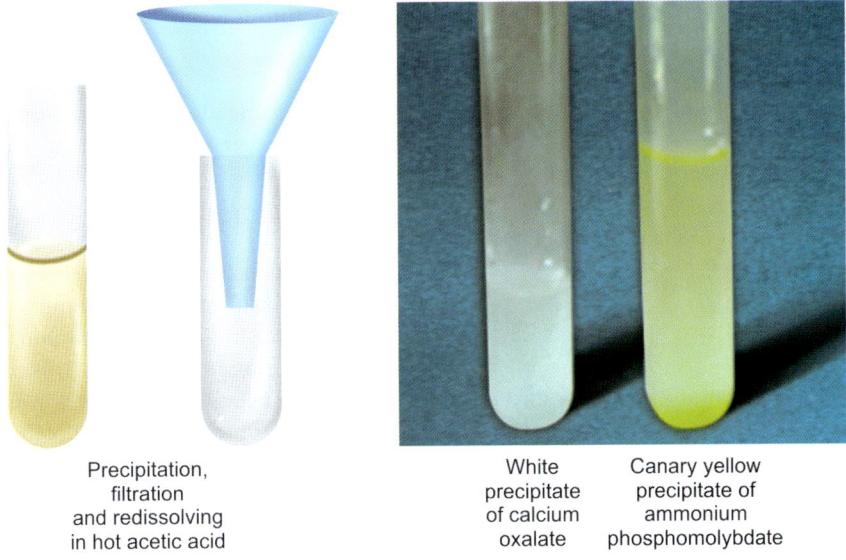

Figure 8.1 Test for calcium and phosphate

lein indicator to show pink color since ammonia is alkaline (Color range of phenolphthalein—colorless to pink; pH range—8.3–10).

Organic Constituents of Urine

Important organic constituents in urine are urea, uric acid, ethereal sulfates, creatinine, organic sulfates, and urinary pigments.

Urea

The amino acids released as a result of protein breakdown are transdeaminated to release ammonia. The toxic ammonia is converted to less toxic urea in the liver. Urea in the blood is sometimes denoted as blood urea nitrogen (BUN). Urea is filtered at the glomerulus and 40–50% of the filtered urea is reabsorbed by the proximal renal tubules.

Causes of high urea content in urine:
- High protein diet
- Conditions leading to increased tissue break-down (increased protein catabolism), e.g. fever, diabetes mellitus, adrenal cortical hyperactivity.

Causes of low urea content in urine:
Liver diseases: In severe diseases of liver, urea synthesis is ↓.

Urea levels in blood and urine in healthy individuals:
- **Blood urea:** 20–40 mg%
- **Urinary urea excretion rate in healthy subjects:** 15–30 g/day.

Uric Acid

It is the catabolite of dietary or tissue purine nucleotides. Plasma level of uric acid is variable and it is higher in males than in females. It is completely filterable and it is reabsorbed at the PCT and secreted at DCT.

Causes of high uric acid content in urine:
- High purine diet
- Conditions where there is increased tissue turn over without any impairment of kidney function—leukemia and other malignancies

- Gout
- Cortisone therapy.

Uric acid levels in blood and urine in healthy individuals:
- Males: 3.6–7.7 mg%
- Females: 2.5–6.8 mg%
- Uric acid content in urine: 300–800 mg/day on an average diet.

Creatinine

The compound creatine phosphate formed in liver, kidneys and pancreas is carried by the blood to other tissues like muscles and brain. About 1–2% of creatine in muscle undergo spontaneous conversion to form cyclical anhydride of creatine, the creatinine. Creatinine is filtered by the glomerulus and a small amount of creatinine is reabsorbed in the proximal convoluted tubule and secreted in the distal convoluted tubule in small amounts. Hence, the measurement of creatinine excretion can be used to assess the glomerular filtration function of the kidneys. The amount of creatinine formed in the body at a point of time depends on age, sex and muscle mass and to a lesser extent on the creatine content in the diet (meat muscle rich in creatine). But daily variations of creatinine levels in blood and excretion in urine are very minimal. Because of this fact, creatinine levels are useful to assess the kidney function. Fasting serum creatinine (i.e. creatinine estimation in fasting sample of blood) can eliminate the influence of diet.

Creatinine level in blood in healthy individuals:
- Males: 0.9 –1.3 mg%
- Females: 0.6 –1.1 mg%.

Creatinine content in urine on an average diet in healthy individuals: 1–2 g/day (nearer to higher limit in males and to lower limit in females).

High creatinine levels in *blood*: Seen in renal failure, and nephritis.

High creatinine levels in *urine*: Seen in myopathies, fever, and muscle injuries.

Organic Sulfates

Urinary sulfates are of three types:
1. **Inorganic sulfates** which come from metabolism of sulfur containing amino acids.
2. **Ethereal or organic sulfates:** This constitutes 10% of total sulfates excreted in urine. The different types of ethereal sulfates seen in human urine are conjugated phenols, phenol sulfuric acid, p-cresol sulfuric acid, skatoxyl sulfuric acid and indoxyl sulfuric acid (indican) (Fig. 8.2). Altogether the excretion rate is **0.04 to 0.1 g per day**. In health the ratio of sum of ethereal and neutral sulfates to inorganic sulfate is about 1: 10.

Formation of ethereal sulfates: Phenols are produced during putrefaction of proteins in the intestine. Phenols reach liver where they are conjugated to form phenol potassium sulfates and excreted as such in urine. Indole and skatole derived from amino acids are oxidized to indoxyl and skatoxyl respectively and conjugated and then excreted in urine. Action of intestinal bacteria on tryptophan leads to the formation of indoxyl sulfuric acid and it is excreted in the urine as potassium salt (indican). Excretion rate of **indican** alone give a rough estimate of intestinal

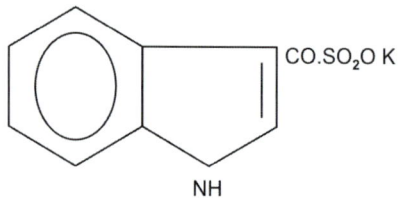

Figure 8.2 Indoxyl sulfuric acid

putrefaction. **Excretion rate of indican in normal individuals 10–20 mg/day.** Its excretory rate **increases** with high meat diet and **decreases** with high carbohydrate diet. **Pathological increase** seen in:

- Intestinal obstruction which causes stagnation of intestinal contents and putrefaction there upon.
- In situations where there is bacterial decomposition of body proteins, e.g. gangrene, putrid pus formation**.**

3. **Neutral sulfates:** It is produced from endogenous sources and its rate of excretion does not change with diet. Generally sulfur in these compounds is in unoxidized or neutral state. The compounds coming under this category are cystine, methyl mercaptan, ethyl sulfide, thiocyanates, taurine derivatives. Neutral sulfur content of normal human urine is 5–25% of total sulfur content **(0.08–0.16 g/day).** Its content in urine is **raised** in cystinuria.

Urobilinogen

After the life span of 120 days, red blood cells undergo lysis and hemoglobin is released in the reticuloendothelial system (spleen, bone marrow, kupffer cells in liver). Hemoglobin splits to form heme and globin. Globin disintegrates into amino acids and heme oxidized by heme oxygenase to form biliverdin, a green pigment which then reduced to a yellow pigment called bilirubin. Bilirubin released into blood is transported by albumin to the liver where it is conjugated with glucuronic acid to form bilirubin glucuronide (conjugated bilirubin) which then passes to the intestine via common bile duct.

In the intestine, the conjugated bilirubin is deconjugated and reduced by intestinal bacteria to urobilinogen and stercobilinogen. The stercobilinogen mainly excreted in feces. Some of the urobilinogen is reabsorbed into the portal circulation and reaches liver. The greater part of this fraction is re-excreted by the liver in the bile as urobilinogen. A small part enters the systemic circulation and is excreted in the urine as "urobilinogen".

Demonstration of Organic Constituents of Urine (Table 8.4)

Test for Urea

- Alkaline hypobromite test
- Specific urease test (Chapter 4–Reactions of Urea).

Test for Uric Acid

- Benedict's uric acid test
- Schiff's test
- Murexide test (Chapter 6–Reactions of Uric acid).

Test for Creatinine

Jaffe's test (Chapter 5–Reactions of Creatinine).

TABLE 8.3 Tests for inorganic constituents in urine

S. no.	Experiment	Observation	Inference
1.	Test for chloride	White precipitate	Precipitate due to silver chloride
2.	Test for sulfate	White precipitate	Precipitate due to barium sulfate
3.	Test for calcium	White precipitate	Precipitate due to calcium oxalate
4.	Test for phosphate	Canary yellow precipitate	Precipitate due to ammonium phosphomolybdate
5.	Test for ammonia	Phenolphthalein at the tip of the glass rod turns pink	Ammonia vapors emerging from the tube turns the phenolphthalein pink

Test for Ethereal Sulfates

Procedure: To 5 mL of urine add 2 mL barium chloride and 2 mL hydrochloric acid. Mix well and filter. Divide the filtrate into two tubes. Boil the contents in one tube. Carefully look for the turbidity developing in the tubes.

Principle: Hot HCl hydrolyzes ethereal sulfate to inorganic sulfate which then gives precipitate of $BaSO_4$ upon reacting with barium chloride.

Test for Urobilinogen: Ehrlich's Test

Procedure: To 5 mL urine, add 1 mL Ehrlich reagent, mix well and keep for 5 minutes. Set a control tube also.

Observation: Red color develops (Fig. 8.10).

Principle: Urobilinogen reacts with p-dimethylaminobenzaldehyde of the reagent to form the red colored complex.

Inference: Normal urine gives a faint red color due to the presence of trace amounts of urobilinogen. It is **excreted increasingly** in urine in hemolytic jaundice where the RBCs are destroyed at a higher rate.

Points to Ponder: Urobilinogen in urine oxidizes to urobilin on keeping and hence stored urine may not answer the test (Fig. 8.3). Fresh urine is preferred for testing urobilinogen.

ANALYSIS OF ABNORMAL CONSTITUENTS OF URINE

Clinical laboratory analysis of urine is useful for diagnosis of several clinical conditions, e.g. diabetes mellitus, phenylketonuria, maple syrup urine disease, alcaptonuria and several others. Urine testing is advantageous as it involves no pain or any disturbance to the patient. Properly collected, analyzed and interpreted urine laboratory tests are valuable for Modern Medicine (Table 8.4).

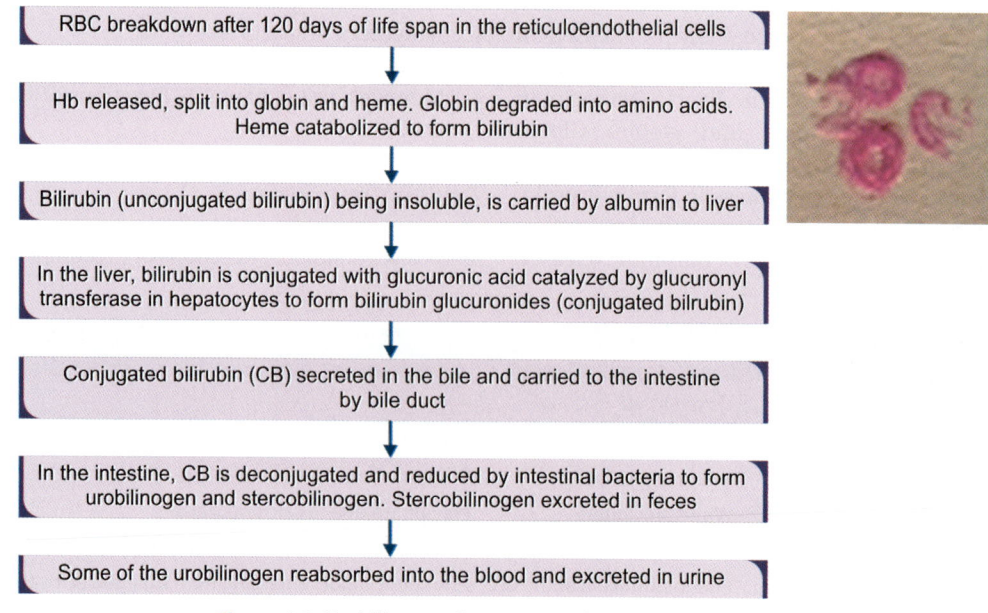

Figure 8.3 Urobilinogen formation and its excretion in urine

TABLE 8.4 Tests for organic constituents

S. no.	Experiment	Observation	Inference
1.	Test for urea: • Alkaline hypobromite test • Specific urease test	• Brisk effervescence • Pink color	• Due to evolution of N_2 gas • Urease enzyme split urea to form ammonia making the medium alkaline. In alkaline medium, phenol red used in the test gives pink color
2.	Test for uric acid • Benedict's uric acid test • Schiff's test	• Intense blue color • Black color	• Phosphotungstic acid reduced by uric acid to tungsten blue • Silver nitrate reduced by uric acid to metallic silver
3.	Test for creatinine • Jaffe test	Orange red color	Due to the formation of creatinine picrate
4.	Test for ethereal sulfate	White precipitate in trace	Precipitate due to barium sulfate
5.	Test for urobilinogen • Ehrlich's test	Red color	Urobilinogen forms complex with para dimethylamino-benzaldehyde to give red color

Glucose

Benedict's Test (Chapter 1–Reactions of Carbohydrates)

Procedure: To 5 mL of Benedict's reagent taken in a test tube, add 8 drops of urine. Shake well and boil for 1 or 2 minutes or keep it in a water bath for 5 minutes.

Observation: A colloidal precipitate forms and the color of which may be green, yellow, orange or red depending on the concentration of sugar in urine.

Interpretation: In the presence of over 0.2–0.3% (0.2–0.3 g/100 mL) of glucose in urine, the precipitate form readily. In the absence of glucose, the solution may remain clear or will show turbidity due to precipitated urates.

Color of the precipitate give an idea about the concentration of the sugar solution as shown below.

Blue	-	absence of reducing sugar
Green	-	up to 0.5 g%
Yellow	-	> 0.5 to 1.0 g%
Orange	-	> 1.0 to 2.0 g%
Brick red	-	> 2 g%

Normally glucose is absent in urine. Appearance of glucose in urine is referred to as glucosuria. **Glucosuria occurs in:**
- Diabetes mellitus
- *Endocrine hyperactivity:* Hyperthyroidism, hyperpituitarism and hyperadrenalism
- *Renal glucosuria:* Here renal threshold for glucose is lowered. Glucose appears in urine even if its level remains within normal limits. This may happen in pregnancy and in inherited lowered renal threshold for glucose. It is differentiated from diabetes mellitus by oral glucose tolerance test (OGTT)
- Alimentary glucosuria.

Points to Ponder

Benedict's test is a nonspecific test since it involves reduction of cupric ions to cuprous ions by any reducing agent. In the urine, several such reducing agents may occur. Such substances are given below.

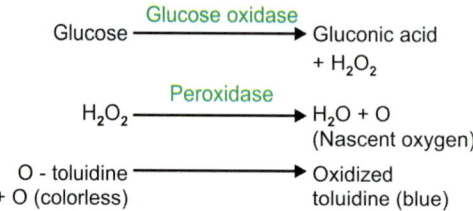

Figure 8.4 Principle of clinistix

Carbohydrate substances: Fructose, galactose, lactose and pentoses.

Noncarbohydrate substances: Ascorbic acid, homogentisic acid.

Presence of glucose can be confirmed by a specific test using glucose oxidase enzyme.

Clinistix: Stiff cellulose strip which turns from red to purple when dipped into urine containing glucose, detect 0.1% (0.1 g/dL) glucose or less. It is more sensitive than Benedict's test. Urine containing low amounts of glucose escapes detection by the Benedict's test (reduction test) but detected by clinistix.

Principle of clinistix: Oxidation of glucose by glucose oxidase to produce gluconic acid and hydrogen peroxide. Hydrogen peroxide acted upon by peroxidase to produce nascent oxygen which in turn acts upon the chromogen (e.g. orthotoluidine) to produce color (Fig. 8.4).

Protein

Heat Coagulation Test (Fig. 8.5)

Procedure: Fill three-fourth of the test tube with urine. Heat the upper one-third of the urine column by a small flame, so that lower two-thirds will serve as control. Add a drop of 30% (v/v) acetic acid to it.

Observation: White turbidity or coagulum.

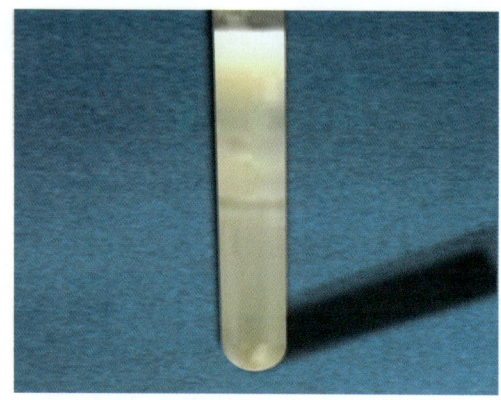

Figure 8.5 Heat coagulation test

Interpretation: White turbidity if disappears on addition of acetic acid, indicates the presence of phosphates or carbonates. If the white turbidity formed remains or appears or intensifies on adding acetic acid points towards the presence of albumin. Addition of acetic acid improves the formation of turbidity since the acidification brings the pH of the medium towards 4.7 (IEP of albumin).

Points to Ponder

- There are chances to miss the presence of albumin in the urine if the pH of urine is high and not lowered by adding acetic acid. Isoelectric point of human albumin is 4.7.
- Normal urine contains less than 30 mg/24 hours and it escape detection by the usually employed methods. Pathologically different proteins detected in urine are albumin, myoglobin, fibrin and oxyhemoglobin.
- The proteinuria is most commonly due to leakage of serum albumin since it is the most abundant and the smallest protein in the serum.
- Albumin most often appears in urine due to altered structure of glomerulus in various kidney diseases.
- Albumin may appear in urine by entering below the kidneys (not by glomerular filtration) from blood, exudates or lymph is called *false albuminuria*.

- **Benign proteinuria:** It is transient and not associated with any kidney disease and occurs with severe exercise and cold bath.
- **Orthostatic albuminuria:** This denotes appearance of albumin in urine after prolonged standing.

Albustix: It is a stiff cellulose strip impregnated at one end with indicator tetrabromophenol blue buffered at pH around 3 which has a yellow color at pH 3.0. Buffer maintains the pH at 3 and hence pH of urine does not interfere. If protein is absent, the color will be yellow. In the presence of protein, the color varies from green to blue.

Points to Ponder
Highly alkaline urine and stale urine (due to the formation of ammonia) may overcome the buffering action of the strip and give a false positive response.

Ketone Bodies

Rothera's Test (Fig. 8.6)

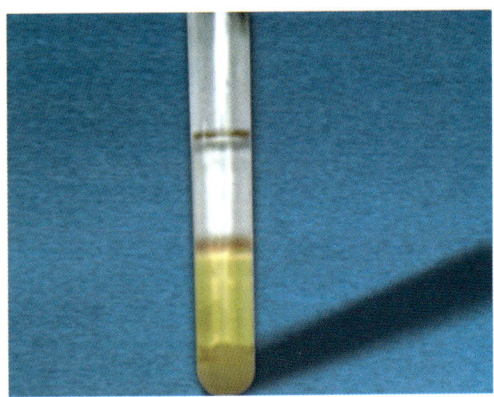

Figure 8.6 Rothera's test

Procedure: Saturate 5 mL of urine with ammonium sulfate crystals and add 2 drops of freshly prepared 2% sodium nitroprusside or a little of sodium nitroprusside powder. Shake well. Add 1 mL of liquor ammonia through the sides of the test tube.

Observation: Reddish violet ring at the junction of two liquids.

Principle: Acetone and acetoacetic acid react with sodium nitroprusside (nitroferricyanide) in the presence of alkali to produce a purple color.

Inference: Indicates the presence of acetoacetic acid and acetone. Normal urine contains ketone bodies approximately 20 mg/24 hours only which is not detectable by usual tests. Ketone bodies are produced excessively in the body in starvation and in uncontrolled diabetes mellitus. Ketone (acetone) bodies include acetone, acetoacetic acid and β hydroxybutyric acid.

Points to Ponder
To detect β hydroxybutyric acid a modified test has to be done. (For this elimination of acetoacetic acid and acetone in the urine sample and oxidation of β hydroxybutyric acid with hydrogen peroxide to form acetoacetic acid is required).

Modified test for β hydroxybutyric acid: Add a few drops of acetic acid to 2 mL of 1:1 diluted urine with distilled water. Boil for few minutes to discard the acetone and the acetoacetic acid present in the urine. Then add 1 mL of hydrogen peroxide and warm gently and carry out Rothera's test. It will give a positive response if β hydroxybutyric acid is present in the urine.

Blood

Benzidine Test

Procedure: Take 2–3 mL of urine in a test tube. Boil for 5 minutes and cool it. Mix equal volumes of benzidine solution (2–3 mL) and hydrogen peroxide in a test tube and add the boiled cooled specimen of urine into the reagent mixture.

Observation: A transient blue color appears.

Inference and Principle: Peroxidase activity of heme oxidizes hydrogen peroxide to release the nascent oxygen which acts upon benzidine to form blue colored compound (Fig. 8.7).

Interpretation: Presence of blood in urine indicates either hematuria (intact RBCs in urine seen in kidney diseases) or hemoglobinura (Hb in urine).

Points to Ponder
- Benzidine is a carcinogen. So, care should be taken while handling the reagent.
- H_2O_2 deteriorates rapidly, so freshly prepared H_2O_2 should be used.
- Boiled cooled urine, must be used for the test otherwise peroxidases of leukocytes present in the urine will act on H_2O_2 to release nascent oxygen to produce false positive reaction. Boiling will help to denature leukocyte peroxidase enzyme.

Strip Test for Detecting Blood or Heme: It is based on the peroxidase activity of heme which splits H_2O_2 to form nascent oxygen which in turn oxidize the chromogen (usual chromogen used is tetramethyl benzidine or orthotoluidine) to form the color.

Upon dipping strip in urine
- Yellow color indicates the absence of heme (either in RBC or in free Hb)
- Blue green color → presence of heme.

Points to Ponder
Ascorbic acid and nitrites interfere with it. Formalin if used as a urinary preservative will also give a false negative test.

Peroxidase activity of heme

$H_2O_2 \rightarrow [O] + H_2O$

Benzidine + [O] → Blue color

Figure 8.7 Principle of benzidine test

Bile Salt

Hay's Test (Fig. 8.9)

Procedure: Take 5 mL of urine in a test tube and sprinkle sulfur powder on the surface of urine.

Observation: Sulfur powder sinks to the bottom.

Inference: Bile salts are present in urine.

Principle: Bile salts reduce the surface tension. Hence, the sulfur powder sinks to the bottom.

Interpretation: Salts of taurocholic acid and glycocholic acid present in the bile regurgitate into blood whenever there is obstruction to bile flow (obstructive jaundice) and will appear in urine. This test is useful to differentiate obstructive jaundice from hemolytic jaundice.

Obstructive jaundice seen with bliary atresia, obstruction of bile duct due to stones or tumors and hepatic jaundice (due to inflammatory intrahepatic cholestasis).

Bile Pigment

Modified Fouchet's Test (Fig. 8.9)

Procedure: To 10 mL urine add 1 mL $MgSo_4$ and boil. While boiling add 10% $BaCl_2$ drop by

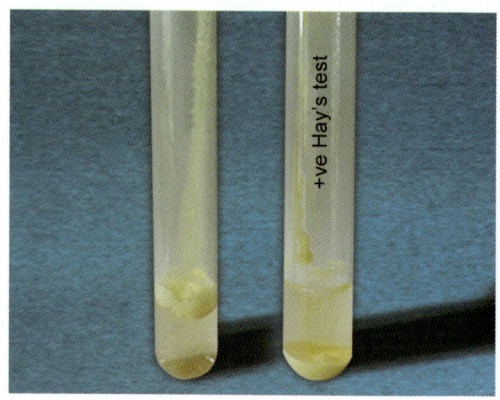

Figure 8.8 Hay's test

drop till maximum precipitate is got. Filter and discard the filtrate. Take the filter paper from funnel and dry the precipitate by mopping with another piece of filter paper. After drying add 2 drops of Fouchet's reagent to the precipitate on the filter paper (Fig. 8.9).

Observation: A bluish green color in the presence of bile pigments.

Principle: Bile pigment present in the urine adsorb onto the $BaSO_4$ precipitate. When Fouchet's reagent (Ferric chloride in trichloroacetic acid) is added, ferric chloride oxidizes bilirubin to bluish green biliverdin and Fe^{3+} (ferric ions) in turn reduced to Fe^{2+} (ferrous ions).

Interpretation: Positive Fouchet's test indicates the presence of conjugated bilirubin in the urine. (Only the conjugated bilirubin can appear in urine in a person with normally functioning kidneys). Conjugated bilirubin appears in urine in cases of **obstructive jaundice** and **obstructive phase of hepatocellular jaundice**. So, this test is useful to differentiate obstructive jaundice from hemolytic jaundice in which it will be negative or weakly positive.

Urobilinogen

Ehrlich's Test

Procedure: Add 1 mL of Ehrlich's reagent (2% paradimethyl aminobenzaldehyde in 20% HCl) to 10 mL of freshly voided urine. Shake well and keep it in the rack for 5 minutes for the color development.

Observation: Normal urine gives only a faint red color.

Principle: Urobilinogen forms a colored adduct with paradimethyl aminobenzaldehyde.

Interpretation: Intensity of red color is related to the concentration of urobilinogen in the following manner (Fig. 8.10).
- **No red color:** Urobilinogen absent.
- **Faint pink color:** Urobilinogen present in normal amounts.
- **Distinctly red color:** Urobilinogen present in increased amounts.

Points to Ponder
Bilirubin if present in the same sample may also react in the same way as that of urobilinogen.

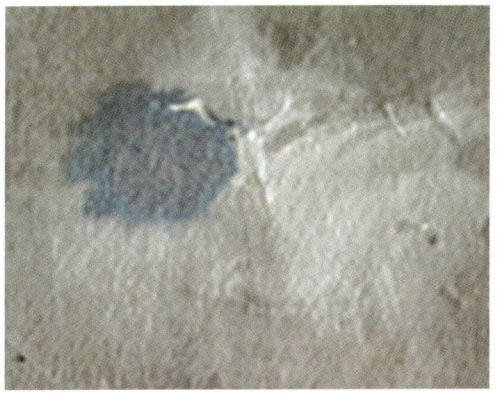

Figure 8.9 Modified Fouchet's test

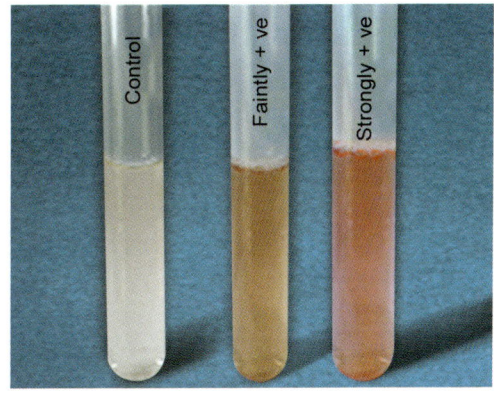

Figure 8.10 Ehrlich's test

In order to avoid this remove bile pigments by adding 2 mL of 10% calcium chloride solution to the urine. Filter and carry out the test with the filtrate.

QUESTIONS

1. Give brief answers:
   a. Significance of odor of urine
   b. Polyuria
   c. Oliguria
   d. Importance of observing the color of urine
   e. Normal pH range of urine
   f. Influence of diet on the pH of urine
   g. Physiological range of specific gravity
   h. Physiological causes of changes in volume of urine
   i. Pathological causes of oliguria and polyuria
   j. Fixed specific gravity
2. Give the excretion rate of the following in urine in a normal person:
   a. Calcium
   b. Phosphates
   c. Sulfates
   d. Chloride
   e. Uric acid
3. Give the principle of the following tests:
   a. Test for chloride
   b. Test for sulfates
   c. Test for calcium
   d. Test for phosphates
4. Give brief answers:
   a. Source of sulfates in urine
   b. Importance of Rothera's test in clinical medicine
   c. Biochemical principle of treatment of urea cycle disorders
   d. Urinary tests useful in the differential diagnosis of jaundice

REAGENT PREPARATION

- **Concentrated Nitric Acid:** Supply from the bottle (16N).
- **3% Silver Nitrate Solution:** Weigh 3 g silver nitrate add to a small volume of distilled water taken in a 100 mL volumetric flask, shake well and make up to 100 mL.
- **10% Barium Chloride Solution:** Weigh 10 g barium chloride add to a small volume of distilled water taken in a 100 mL volumetric flask, shake well and make up to 100 mL.
- **2% Potassium Oxalate:** Weigh 2 g potassium oxalate add to a small volume of distilled water taken in a 100 mL volumetric flask, shake well and make up to 100 mL.
- **Ammonium Molybdate Solution:** Dissolve 100 g of molybdic acid in 144 mL of ammonium hydroxide (specific gravity 0.90) and 271 mL water. Add 489 mL nitric acid (specific gravity 1.42) to this solution slowly with constant stirring and add 1148 mL water. Keep the mixture in warm place for several days. Check the adequacy of keeping in the following manner. Take about 5 mL of this solution in a tube and heat up to 40°C and if no yellow precipitate of ammonium phosphomolybdate is forming, it can be considered fit for use.
- **Urease Solution:** Grind 10 g horse gram or jackfruit seeds (rich sources of urease enzyme) with 100 mL of 30% ethanol using a mortar and pestle and filter.
- **Fouchet's Reagent:** Dissolve 25 g trichloroacetic acid in about 50 mL of water and add 10 mL of 10% ferric chloride and make up to 100 mL with water.
- **Ehrlich's Reagent:** Dissolve 2 g p-dimethylaminobenzaldehyde in 100 mL of 20% hydrochloric acid.
- **Phenolphthalein 0.1% (pH range–8.3–10 color range—colorless to red):** Dissolve 0.1 g Phenolphthalein in 100 mL of 50% alcohol.

REFERENCE

1. Klahr S. Structure and function of the kidney. In: Cecil Textbook of Medicine, 17th ed. JB Wyngaarden, LH Smith Jr (Eds), Philadelphia, WB Saunders Co;1985.

CHAPTER

9

Identification of Hemoglobin Derivatives

SPECTROSCOPIC EXAMINATION OF HEMOGLOBIN PIGMENTS

Spectroscopy and Spectroscope

Spectroscopy involves observation and study of absorbed light by means of a spectroscope. Spectroscope is an instrument used to study the absorption spectra of various substances. It consists of a prism that refracts the light or gratings for diffraction of light and an arrangement for rendering the rays parallel and a telescope that magnifies a spectrum.

Many biological substances have characteristic light absorbing properties because many molecular groupings have characteristic light absorption pattern which can be used for detection and quantitative assay. These substances can be identified by studying their light absorption properties, e.g. hemoglobin, porphyrins.

Spectroscope (Fig. 9.1) consists of a prism situated behind an adjustable eyepiece. When a beam of light is allowed to pass through the prism, it disperses light into solar spectrum (Fig. 9.2). When a colored solution is inter placed in the path of the light entering the prism through the slit of the spectroscope, it will alter the appearance of the spectrum. If the solution absorbs light of particular wavelength correspondingly dark bands will be produced in the spectrum. It is possible to identify specific substances by accurately positioning the absorption bands produced in the spectrum by the solution.

Various types of spectroscopes are available. For the identification of Hemoglobin (Hb) derivatives or porphyrins in biological fluids, a simple hand-held spectroscope is used.

In the solar spectrum, several vertical dark lines, called Fraunhofer lines are seen. These dark lines are caused by the absorption of the white light from the hotter regions of the sun by chemical elements present in the cooler parts of the sun. They are designated as A, B, C, D, E—so on—from the red end of the spectrum.

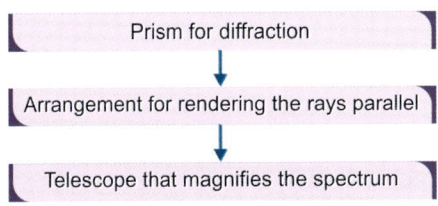

Figure 9.1 Essential parts of a spectroscope

Photosphere is the visible, intensely luminous portion of the sun which has an estimated temperature of 6,000 K.

Chromosphere is a layer of sun's atmosphere surrounding the photosphere which is visible during a total eclipse. The chromosphere is several miles thick and has an estimated temperature of 20,000 K.

The spectroscope consists of a long tube (slit tube) and a short tube (tube holder). Long tube contains the optical system and the short tube contains the wavelength scale. Adjust the slit in the long tube by adjusting the knurled ring. Slit should be very narrow. Then the instrument is directed to sunlight. When it is viewed through eyepiece, the spectrum is visible. By adjusting the slit tube vertical narrow absorption bands can be viewed in the spectrum. The wavelength scale can be adjusted using the tube holder, so that graduations in the scale appear in short focus. Graduations are done in nanometers (nm). Set the scale with reference to D line (i.e. 590 nm marking should correspond to D line in the spectrum). D line is the most important of the Fraunhofer lines. D line corresponds to absorption band of sodium with a wavelength of 590 nm. This marking is extended downwards. Deviations in the D line can be adjusted using the adjusting screw.

Principle: A beam of sunlight is dispersed into 7 components (VIBGYOR) using a prism incorporated in the spectroscope. When a tube containing a Hb solution, is kept against the spectroscope, the Hb pigment in solution will absorb a certain portion of the visible light and that will be seen as dark bands in the spectrum when viewed through the spectroscope. The bands produced will be characteristic of the molecular nature of the substance.

Procedure: Set the spectroscope for reading. Take the solution to be examined in a clean dry test tube and hold it vertically in front of the instrument. View the absorption bands produced and note their position on the wavelength scale incorporated.

Precautions and General Rules

- Spectroscope should be adjusted properly.
- Bright light should be used.
- Before colored pigments are examined, it has to be diluted in the right concentration. In higher concentrations, the bands are darker and broader. In low concentrations, the bands are narrow and sharp.
- Bands when viewed on spectroscopic examination are read from the red end of the spectrum as α, β, γ and so on.

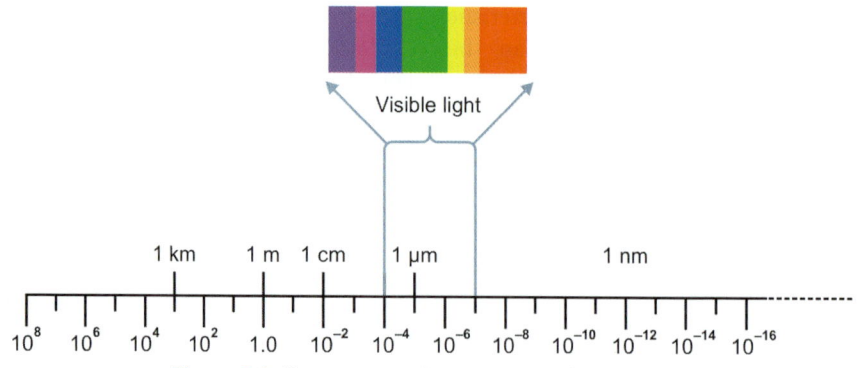

Figure 9.2 Electromagnetic spectrum and visible light

- Position of bands is read from the scale. The scale reading against the midpoint of the band is read.

Identification of Different Hb Derivatives (Fig. 9.3)

Oxy Hemoglobin

Two atoms of oxygen are taken up by each atom of iron. One iron atom is attached to each monomer of Hb. So, a tetramer Hb molecule can take up 4 molecules of oxygen. One gram of Hb can combine with 1.36 mL of oxygen at NTP. Oxy-Hb loses oxygen when exposed to a low oxygen pressure or upon treating with ammonium sulfide or sodium dithionite ($Na_2S_2O_4$).

Color: Orange red
- **Spectroscopy:** 2 bands seen. α band at 577 nm in the yellow region and β band at 541 nm in the green region and γ band at 413 nm in the violet region. Alpha band is narrow and sharp and beta band is broad and hazy. Generally alpha and beta bands are looked for as the gamma band will not be distinguishable.
- **Schumm's Test:**
  - *Principle:* Involves reduction of oxy-Hb to deoxy-Hb by the action of reducing substances.
  - *Procedure:* Take 2 mL of the specimen containing oxy-Hb in test tube. Add 1 mL of ether to cover the surface in order to avoid contact with air and then add 0.2 mL of ammonium sulfide solution. Examine spectroscopically.
  - *Observation:* Alpha band disappears.
- **Addition of sodium dithionite:**
  - *Principle:* Involves reduction of oxy-Hb to deoxy-Hb by the action of reducing substances.
  - *Procedure:* Take 2 mL of the specimen containing oxy-Hb in a test tube and add a pinch of sodium dithionite.

Note the color change and examine spectroscopically.
- *Observation:* Orange red colored oxy-Hb changes into purple colored deoxy-Hb.
- *Clinical application:* In the diagnosis of *intravascular hemolysis.*

Hemoglobin (Hb) released from the cells is bound to haptoglobin (a glycoprotein) and haptoglobin Hb complex is rapidly removed by reticulo- endothelial system. When the binding capacity of the haptoglobin is exceeded (40–160 mg of Hb/dL), free Hb accumulates in the plasma. Renal threshold of free Hb is 0.15 g/dL. When the concentration of free Hb in the plasma exceeds 0.15g/dL, free Hb starts appearing in urine. In such cases, plasma and urine shows the presence of oxy-Hb.

Deoxyhemoglobin (Reduced Hb)

Formed by deoxygenation of Hb.

Color: Purple
- **Spectroscopy:** A single broad band. The mid-point of this band corresponds to 565 nm in the green region.
- **Upon shaking vigorously** the tube containing reduced Hb solution forms oxy-Hb.
  It is obvious by the change in color from purple to orange red and on spectroscopic examination gives the bands of oxy-Hb.
- **Carboxy-Hb (Carboxyl Hb):** Formed by the action of carbon monoxide on hemoglobin and oxy-hemoglobin. CO has high affinity towards Hb than O_2. So, it can displace oxygen from oxy-Hb to form carbonyl Hb.

Color: Pink
- **Spectroscopy:** 2 bands seen. α band at 572 nm in the yellow region and β band at 535 nm in the green region and γ band at 418 nm in the violet region. Alpha band is narrow and sharp and beta band is broad and hazy. Generally alpha and beta bands

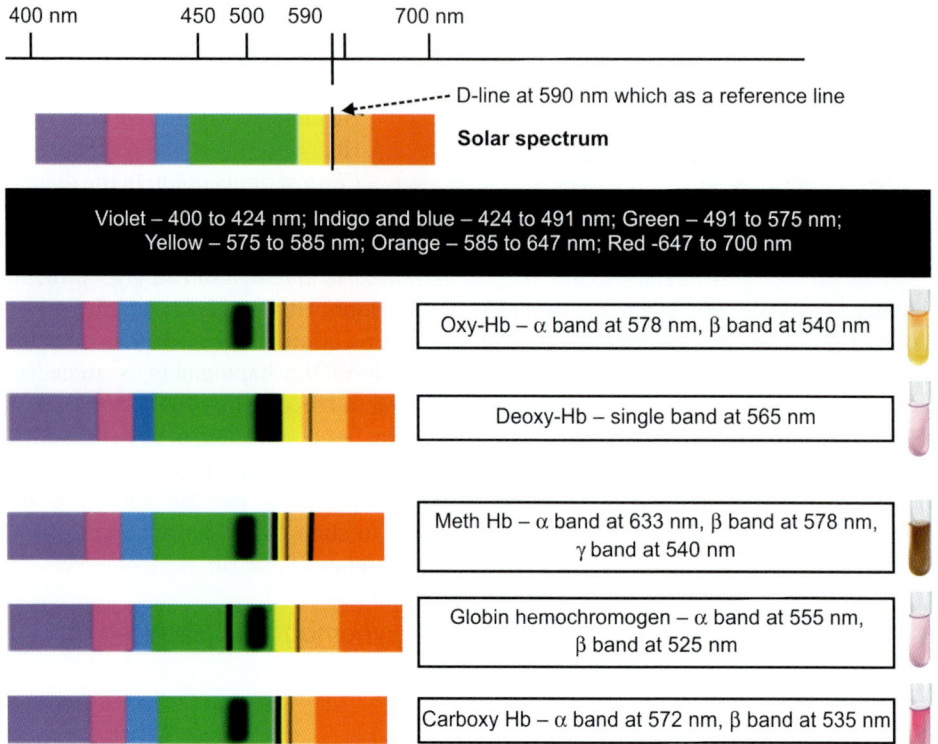

Figure 9.3 Identification of Hb derivatives by spectroscopy

are looked for as the gamma band will not be distinguishable (Fig. 9.3). The bands look very much similar to oxy-Hb but can be differentiated from it by the following tests.
- **Addition of reducing agents** like sodium dithionite and ammonium sulfide to carbonyl Hb solution will not produce any effect whereas oxy-Hb will be converted into deoxy-Hb as indicated by the color change (purple color for deoxy-Hb, orange red color for oxy-Hb). But heat and strong reducing agents can convert carbonyl Hb into hemoglobin.
- **Addition of alkalies:** Mix equal volumes of Hb solution and 25% sodium hydroxide. Oxy-Hb turns brown but carboxy-Hb remains as such.

Points to Ponder
- Blood collected from patients suspected of CO poisoning, should be taken under oil in order to avoid contact with atmospheric oxygen or by vaccutainers.
- Carboxy-Hb being photolabile, must be collected in amber colored bottles.

Methemoglobin

The iron in meth-Hb is in the ferric state and hence it cannot transport and deliver oxygen. Normally only 1% of Hb exist as meth-Hb. Meth-Hb reductase reduce meth-Hb formed in the cells to Hb and there by only low levels of meth-Hb is seen in the tissues (maximum of 1% of the total Hb). Its concentration in the body fluids may be raised (meth-hemoglobinemia) in congenital meth-hemoglobinemia and in dapsone poisoning. Meth-Hb is induced by cerain drugs like phenacetin, sulfonamides, primaquine. Treatment is by giving reducing agents like methylene blue, vitamin C, etc.

If the red cells containing meth-Hb undergo hemolysis, meth-Hb will be liberated and can be filtered by the glomerulus and reaches urine. In such situations, meth-Hb may be present in tissues, plasma and urine.

Color: Reddish brown

- **Spectroscopy:** 3 bands are seen. The characteristic alpha band in the orange region at 633 nm, beta band at 578 nm in the green region and gamma band at 540 nm in the green region.
- **Addition of reducing agent, the sodium dithionite:** This reduces meth-Hb and Oxy-Hb to deoxy-Hb. But carboxy-Hb and sulph-Hb are not affected.
- **Addition of alkali:** Add 2 drops of ammonia solution to 5 mL of Meth-Hb solution. This causes conversion of meth-Hb to alkaline meth-Hb. Alkaline meth-Hb does not produce a band at 633 nm.

Globin Hemochromogen (Heme + Denatured Globin)

This is an important derivative of Hb. It absorbs light and gives visible bands even in high dilutions. So, it is useful to test the presence of traces of blood in body fluids in clinical practice and to aid in the detection and confirmation of suspected blood stains in forensic medicine. For this purpose, the sample (1 in 100 dilution) should be treated with 3 drops of 5% NaOH. Then heat the solution. It will give yellow solution due to the formation of alkali hematin (hematin = ferric iron + protoporphyrin; heme = ferrous iron + protoporphyrin). Upon addition of alkali, meth-Hb is formed which then decomposes into globin and hematin. Cool and add a pinch of reducing agent like sodium dithionite with gentle shaking in order to obtain pink colored globin hemochromogen (reducing agent reduces hematin to heme to give heme + denatured globin = globin hemochromogen).

Color: Pink

- **Spectroscopy:** Two characteristic bands are seen. Alpha band at 555 nm in the green region and beta band at 525 nm in the green region itself. Unlike oxy-Hb alpha band of globin hemochromogen is more intense and the beta band is faint.

Clinical application: Fetal Hb is relatively resistant to alkali, so it will help to distinguish blood rich in HbF from blood containing predominantly adult hemoglobin (Hb A).

IDENTIFICATION OF HEMOGLOBIN PIGMENT BY MICROSCOPY

Hemoglobin (Hb) pigment can be identified by preparing hemin crystals.

Hemin Crystals

Hemin → chloride of hematin
Hematin → ferric iron + protoporphyrin

Preparation of hemin crystals: Place a drop of blood on a clean glass slide and make a smear on the slide. Place a cover slip over that. Add one drop of Nippe's fluid at one side of the cover slip so that fluid permeates through the undersurface of the cover slip by capillary action. Show the slide over the flame so that the fluid under the cover slip evaporates almost completely. Look under low power and high power of microscope. The hemin crystals will be seen as **brown colored rhombic shaped crystals** (Fig. 9.4).

Principle: Nippe's fluid contains KCl, KBr and KI and glacial acetic acid. Upon heating with

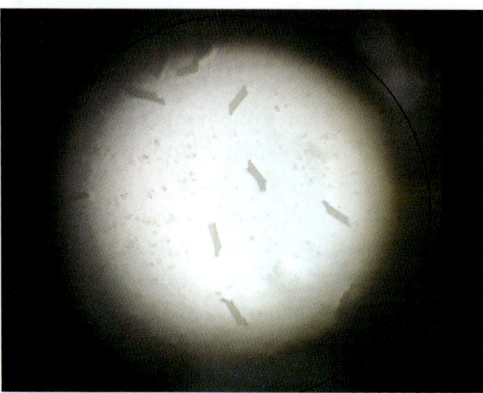

Figure 9.4 High power view of hemin crystals

Nippe's fluid globin is denatured and heme is oxidized to hematin which is then converted into hematin chloride. Hematin chloride is otherwise known as Hemin.

Clinical application: To differentiate between blood stain from other stains.

QUESTIONS

1. Name the following:
   a. Hemoglobin derivative containing ferric iron.
   b. Raised Hb derivative in blood due to dapsone poisoning.
   c. Suffocation due to carbon monoxide is because of the formation of—Hb derivative in blood.
   d. Which is the ionic state of iron in oxy-Hb, Meth-Hb and in globin hemochromogen.
   e. What will happen to oxy-Hb when it is exposed to reducing substances?
   f. One hemoglobinopathy producing meth-hemoglobinemia.
   g. Name two drugs that may cause meth-hemoglobinemia.
   h. Hemoglobin (Hb) derivative imparting dark brown color to blood.
2. How will you identify the following Hb derivatives:
   a. Oxy-Hb
   b. Deoxy-Hb
   c. Carboxy-Hb
   d. Meth-Hb
   e. Globin hemochromogen.
3. Give short answers:
   a. Relevance of identification of globin hemochromogen in clinical and forensic practice.
   b. What are the precautions to be taken while taking blood samples for carbonyl Hb screening?
   c. What is the mechanism of the formation of meth-Hb formation?
   d. What are Fraunhofer lines? Which Fraunhofer line is used as the reference line in the identification of Hb derivatives?
   e. How will you prepare hemin crystals from suspected blood spot from the crime site?
   f. Draw the high power view of hemin crystals.
   g. What is the importance of hemin crystals in forensic medicine?

REAGENT PREPARATION

- **Anticoagulated blood:** Use EDTA or Heparin as anticoagulant. Anticoagulated blood is required for preparing different Hb derivatives.
- **Oxyhemoglobin:** Add 2 mL of blood to 100 mL of distilled water in a conical flask and mix well till a clear solution is obtained. 3–5 mL of this oxy-Hb solution may be taken in test tubes for spectroscopic examination.
- **Deoxy-Hb solution:** Prepare oxy-Hb solution as detailed above and to this add in small quantities (a pinch) of sodium dithonite into the conical flask till the orange red color transform into a purple color.

- **Meth-hemoglobin solution:** Prepare 1 in 50 dilution of blood and take 100 mL of it in a conical flask. Add a pinch of potassium ferricyanide crystals into the conical flask and stir well so as to obtain a black tea colored solution.
- **Globin hemochromogen:** Prepare 1 in 100 diution of blood and take 100 mL of it in a conical flask. Add 3-4 mL 5% NaOH solution to it, so as to get a brown solution and heat gently till the brown color becomes yellowish brown due to the formation of alkali hematin. Cool and add a little sodium dithionite with simultaneous gentle shaking to get pinkish globin hemochromogen solution.

C H A P T E R 10

Reactions of Milk

INTRODUCTION

Milk is secreted by mammary gland during lactation. Under the influence of high circulating levels of estrogen, progesterone, prolactin and perhaps hCG (human chorionic gonodotropin) during pregnancy. After expulsion of placenta, levels of estrogen and progesterone declines abruptly. The sudden drop in estrogen initiates lactation. Prolactin also is involved in producing milk. Major components of milk are casein, lactalbumin, lactoglobulin, lactose, fat, electrolytes (sodium, potassium, chloride, calcium, magnesium, phosphorus and very little amounts of iron) and vitamins (vitamin A, D, B_1, B_2, niacin) vitamin C is deficient in milk.

PRECIPITATION OF CASEIN FROM MILK

Take 20 mL of milk in a 100 mL conical flask or beaker. Add an equal volume of water and 2% acetic acid in drops till a maximum precipitate is obtained. Stir well with a glass rod to break up the precipitate. Keep for 5 minutes and filter through a Whatman No: 1 filter paper. Collect the precipitate containing casein and dry between folds of filter paper. Filtrate contains lactose and minerals. Acetic acid is added in the above procedure to lower the pH to the isoelectric pH of casein (4.6).

Tests with the Precipitate

- **Test for fats: Grease Spot Test:** Add a portion of the precipitate to 3 mL of ether and perform grease spot test (*See* Chapter 3–Reactions of Lipids).

Tests with the Filtrate

- **Test for lactalbumin and lactoglobulin—heat coagulation test:**
    - **Procedure:** Take 10 mL of filtrate in a test tube. Heat the top one-third over a flame.

- **Observation:** A fine coagulum forms.
- **Principle:** Lactalbumin and lactoglobulin present in the filtrate are heat coagulable.
- **Test for lactose:** Do Benedict's test and osazone test.
- **Test for calcium and phosphorous (Fig. 10.1):**
  - *Procedure:* To 10 mL filtrate add 5 drops of concentrated ammonia and heat to boil. Cool. A gelatinous precipitate forms. Filter and discard the filtrate. Add 5 mL warm 16 % acetic acid through the sides of the filter paper to dissolve the precipitate. Divide the solution into 2 portions in two test tubes. To one, add 2 mL of 2% potassium oxalate. To the second tube add 1 mL concentrated HNO_3 and 3 mL ammonium molybdate reagent. Warm the solution.

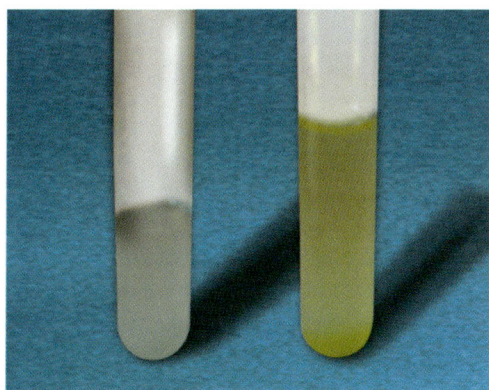

Figure 10.1 Test for calcium and phosphorous

- *Observation:* A white precipitate forms in the first tube and a canary yellow color in the second tube.
- *Principle:* White precipitate due to calcium oxalate formation and canary yellow color due to ammonium phosphomolybdate.

QUESTIONS

1. Name the hormones and mention their role in the secretion of milk.
2. Which are the proteins and carbohydrates of the milk?
3. Name the minerals present in the milk.
4. Name one trace element which is present scarcely in the milk.
5. Mention 4 vitamins present in the milk.
6. How will you precipitate out casein from milk?
7. How will you demonstrate lactalbumin and lactoglobulin in the milk?
8. How can you show that lactose is present in the milk?
9. Describe the experiment to test the presence of calcium and phosphorous in the milk.

REAGENT PREPARATION

- **Concentrated ammonia:** Dispense from the 15 N liquor ammonia bottle available.
- **2% potassium oxalate:** Weigh 2 g potassium oxalate and add to a small volume of distilled water taken in a 100 mL volumetric flask, shake well and make up to 100 mL.

SECTION 2

Quantitative Analysis

CHAPTER 11

Introduction to Quantitative Analysis

PRINCIPLES OF COLORIMETRY

In many diseases, blood levels or urinary excretion values of certain constituents may be varied from normal as a manifestation of the diseased condition. Such changes if measured quantitatively can be used in diagnostic and prognostic purposes. By normal value, it is meant the amount of a particular constituent present in the body fluids of clinically healthy individuals. The normal value is affected by a number of factors such as age sex, type of the individual genetic makeup, etc. Hence, we usually use the term **reference ranges.** These investigations or biochemical assays, which help to reach at a clinical diagnosis constitute what is meant by **clinical chemistry.**

Usually the samples used for routine analysis in a disease are blood, urine, CSF, pleural and peritoneal fluid. Careful attention is needed in all stages of investigation from the collection of samples to the estimation of constituents to avoid false results. Interpretation of results requires information regarding complaints, management and the conditions under which samples are collected and the method of assay.

Collection of Blood Specimen

Capillary or venous blood is taken for almost all the determinations made on blood.

Whole blood is used for determination of pH, analysis of blood gases and determination of concentration of Hb or its derivatives. But majority of the clinical investigations are carried out using *serum or plasma*.

Serum

For separating the serum, blood is collected in clean dry containers and allowed to clot at room temperature. After sometime, the clot retracts and the serum separates as a supernatant layer. This supernatant layer is removed carefully using capillary pipettes into dry bottles. Serum has the advantage that no anticoagulant is necessary but it takes time for the separation of serum. During this period, CO_2 is lost and K^+ and phosphates move from cells to serum. Hence, for electrolyte estimations, plasma is preferred (immediate separation has to be done).

Plasma

To prepare plasma, collect the blood in a container with anticoagulant and then centrifuge. The cells settle down and the plasma separates. The difference between the plasma and the serum is that the serum is devoid of fibrinogen and other clotting factors whereas plasma contains fibrinogen.

Anticoagulants usually used are heparin, potassium oxalate and sodium citrate. Specimen for glucose estimation must contain specific antiglycolytic preservatives such as sodium fluoride (inhibits enolase enzyme thereby inhibiting glycolysis).

Hemolyzed sample when used for estimation will give wrong values. Therefore, during collection of sample, care should be taken to prevent hemolysis. It is better to carry out the investigations immediately after collecting the sample, especially in the case of enzyme assays. If the analysis cannot be carried out immediately, serum has to be preserved in cold. At 4°C it is stable up to a week. But if it is kept frozen the constituents are stable for a longer period.

Sera for bilirubin and calcium estimations need special precautions. Bilirubin is rapidly destroyed by light (photolabile) and hence should be protected from light by keeping in amber colored bottles. Serum for calcium estimations should be kept in glass containers and preserved in refrigerators.

Collection of Urine Sample

Single specimen of urine is used for qualitative tests. But for quantitative tests, 24-hour urine samples are required. 24-hour urine samples are collected in bottles containing suitable preservatives, e.g. concentrated HCl, thymol, toluene, etc. Refrigeration is the best method of urine preservation.

Colorimetry

Colorimetry means measurement based on the intensity of the color. Clinical chemistry estimations are usually carried out by colorimetric procedures. Many substances of biological and medical interests are colored. Colorless constituents can be converted to colored derivatives or colored complexes by subjecting them to undergo specific reactions. If a colored substance present in a solution absorbs light in the visible region of the spectrum, the amount of light absorbed depends on the intensity of the color which in turn influenced by the concentration of the substance in solution.

Quantitative estimation of substances by measurement of the intensity of their colored solutions is known as colorimetric analysis or colorimetry. In this procedure, the color intensity of the solution is compared with that of a standard solution. For this purpose, colorimeters, the instruments that match the colors of unknown and standard solutions are used.

These are of two types:
1. Visual colorimetry
2. Photoelectric colorimeters

In visual colorimeters, the colored solution of an unknown concentration is compared with a standard solution of identical color.

Photoelectric Colorimetry

It is the widely used method for determining the concentration of biochemical compounds. This utilizes the property that when white light passes through a colored solution, some wavelengths are absorbed more than the others. Many compounds are not colored themselves but can

be made to absorb light in the visible region by reaction with suitable reagents. These reactions are specific and sensitive enough to measure compounds present in low concentration like millimoles per liter.
- The depth of the color is proportional to the concentration of the compound being measured.
- Amount of light absorbed is directly proportional to the intensity of the color and hence concentration of the substance. Figure 11.1 gives diagrammatic representation of photoelectric colorimeter.

Beer-Lambert Law is a combined law of light absorption practiced in carrying out photometric assays.

When a ray of monochromatic light of initial intensity 'I_0' passes through a solution in a transparent vessel, some of the light is absorbed so that the intensity of the transmitted light 'I' is less than 'I_0'. There is some loss of intensity of light due to scattering by particles in solution and reflection at the interfaces but mainly by absorption by the solution.

The relationship between the I and I_0 depends on:
- The path length of the absorbing medium 'l'
- The concentration of the absorbing solution 'C'.

These two factors are related in the laws of Lambert and Beer (Fig. 11.2).

1. **Lambert's law:** When monochromatic light passes through a solution, the intensity of light transmitted decreases exponentially with increasing path length.
$$I = I_0 e^{-K_1 l}$$
2. **Beer's law:** When a ray of monochromatic light passes through a solution, the intensity of light transmitted decreases exponentially as the concentration of the medium increases
$$I = I_0 e^{-K_2 c}$$
These two laws are combined together in Beer-Lambert law.
$$I = I_0 e^{-K_3 c l}$$

Transmittance
The ratio of intensities is known as the transmittance (T) and this is usually expressed as a percentage.

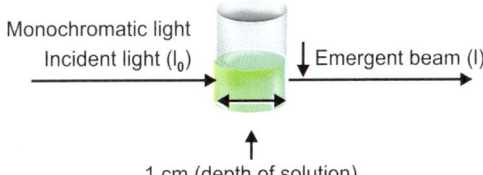

Figure 11.2 Incident and emergent light

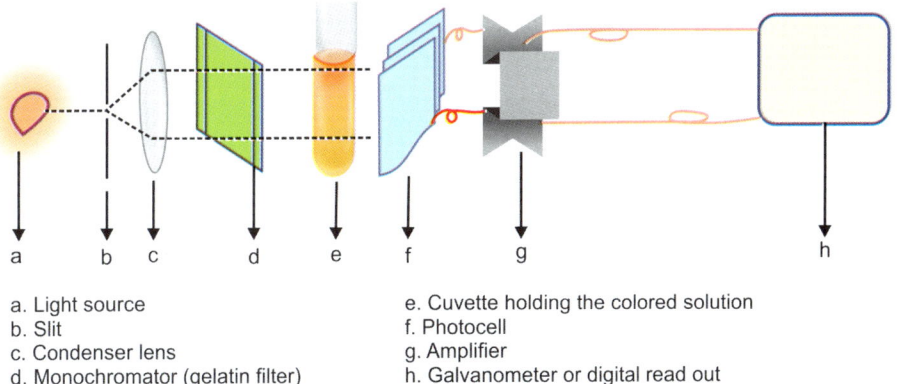

a. Light source
b. Slit
c. Condenser lens
d. Monochromator (gelatin filter)
e. Cuvette holding the colored solution
f. Photocell
g. Amplifier
h. Galvanometer or digital read out

Figure 11.1 Diagrammatic representation of photoelectric colorimeter

Percent transmittance = $I/I_0 \times 100 = e^{-K_3 c l}$

This is not very convenient, which is evident from the negative exponential curve obtained by plotting percent transmittance against concentration (Fig. 11.3A).

Extinction

If logarithms are taken instead of a ratio, then
$$\log_e I_0/I = K_3 c l$$
$$\log_{10} I_0/I = K_3 c l / 2.303$$

The expression $\log_{10} I_0/I$ is known as extinction (E) or absorbance (A). The extinction is sometimes referred to as the optical density. But this term is not recommended.

Therefore, E = Kcl.

If Beer, Lambert's law is obeyed and 'l' is kept constant then a plot of extinction against concentration gives a straight line passing through the origin (Fig. 11.3B) which is convenient than the curve for transmittance.

Some colorimeters and spectrophotometers have two scales—a linear one of percent transmittance and a logarithmic one of extinction (Fig. 11.4).

The extinction scale is linearly related to concentration and is used in the construction of standard curve. With the aid of such a standard curve, the concentration of an unknown solution can easily be determined from its extinction.

Molar Extinction Coefficient

E = kcl

If l = 1 cm ; c = 1 mol/L

E = K

K = molar extinction coefficient

The molar extinction coefficient is characteristic for a compound. Molar extinction coefficient 'K' is thus, the extinction given by 1 mol/L in a light path of 1 cm and is written as E/1 mol/1 cm.

Specific Extinction Coefficient

Molecular weights of some compounds like proteins and nucleic acids in a mixture are not available and in such cases specific extinction coefficient is used. This is the extinction of 10 g/L of the compound in the light path of 1 cm (E/10 g per liter/1 cm).

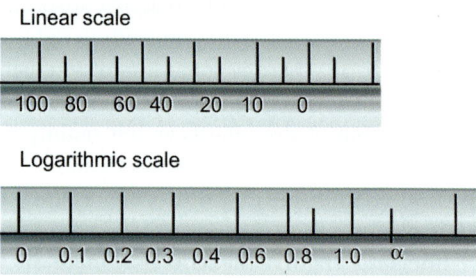

Figure 11.4 Linear and logarithmic scale

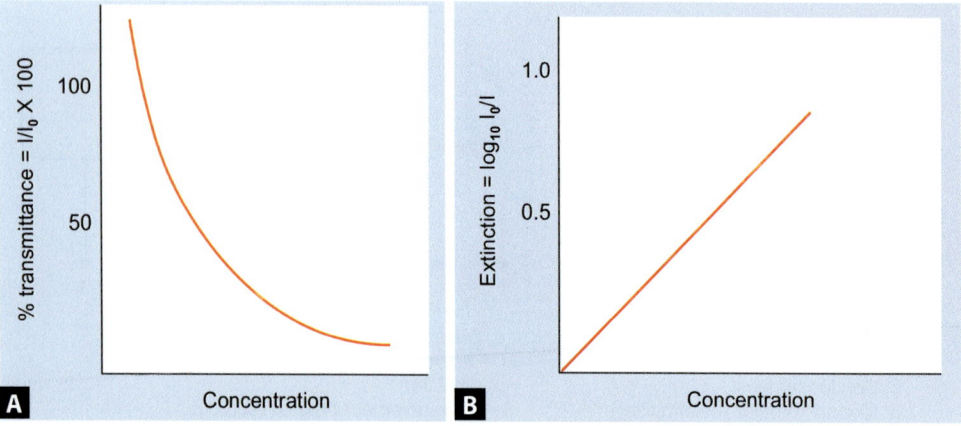

Figures 11.3A and B (A) Percent transmittance vs concentration; (B) Absorbance vs concentration

Limitations of Beer-Lambert Law
Sometimes a nonlinear plot is obtained of extinction against concentration and it is due to one of the following conditions not being fulfilled:
- Light must be of narrow wavelength range and preferably monochromatic.
- Wavelength of the light used should be at the absorption maximum of the solution which gives greatest sensitivity.
- There must be no ionization, association, dissociation or solvation of the solute with concentration or time.
- The solution is too concentrated giving an intense color. The law only holds up to a threshold maximum concentration for a substance.

COMPONENTS OF PHOTOELECTRIC COLORIMETER (FIG. 11.1)

- **Light source:** Usually a tungsten lamp to provide light over a range 200–700 nm.
- **Absorption filters:** These absorb selective wavelengths. Gelatin filters offer a range of alternate wavelengths.
  Examples:
  Blue filter: Provide a light ray around 425 nm.
  Green filter: Provide a light ray around 525 nm.
  Red filter: Provide a light ray around 690 nm.
- **Sample holder or cuvette:** These are glass test tubes made of scratch proof glass. It should be dirt free and of uniform diameter.
- **Photocell:** This is a device which convert light energy to electrical impulse. Light passing through the cuvette containing the solution falls on the surface of a sensitive photocell and this photocell convert light energy to electrical impulse. This electrical impulse is proportional to the emergent light.
- **Galvanometer:** The photocell is connected to a galvanometer whose scale is graduated in percent transmittance or as optical density (OD) units.

MEASUREMENT IN A PHOTOELECTRIC COLORIMETER

In photoelectric colorimeter, absorbance of a substance is found out by measuring the percentage of incident light that is transmitted by the solution.

Percent transmittance = Intensity of emergent light/Intensity of incident light × 100

A more satisfying way of expressing percent transmittance is by optical density.

$OD = -\log T$

The emergent light rays are passed through a photocell which will convert the light energy into electrical impulses. The current thus generated is measured by galvanometer.

Beer's law states that optical density is directly proportional to the concentration. Thus, if a graph is plotted with concentration on X axis and optical density on Y axis a straight line will be obtained and the concentration could be directly read.

When Beer's law is obeyed,

$$\frac{\text{Concentration of unknown}}{\text{Concentration of standard}} = \frac{\text{OD of unknown}}{\text{OD of standard}}$$

∴ Concentration of unknown = OD of unknown/OD of standard × concentration of standard

This is the **principle** of all colorimetric reactions. To measure the concentration of a substance in test solution, three solutions have to be prepared.
1. Test solution which is to be analyzed.
2. Standard solution prepared from known quantity of the substance to be estimated.
3. Reagent blank containing all the reagents but without the substance to be estimated.

Reagent blank compensates for nonspecific color produced by color of reagents or impurities present in the reagent. In practice, the instrument is first set at zero reading by using reagent blank.

Precautions: It is important that the solution should not be cloudy or turbid and should not contain any bubbles since all these will affect percent transmittance.

Selection of Filter

It depends on the color of the solution to be tested. Filter with complement color is used.
Examples:

Color of the solution	Color of filter
Purple	Green (505–555)
Bluish green	Red (650–700)
Yellow	Blue (430–475)

QUESTIONS

1. What is colorimetry?
2. Mention two types of colorimetry.
3. Give the principle of photoelectric colorimetry.
4. State Beer –Lambert law.
5. What do you mean by extinction coefficient (absorbance) of a substance?
6. Mention the parts of a photoelectric colorimeter.
7. What is meant by monochromatic light?
8. Why extinction is preferred than transmittance in the photoelectric colorimetric techniques employed for estimating the concentration of various substances?
9. What are the limitations of Beer-Lambert law?
10. What is reagent blank?

CHAPTER 12

Determination of Glucose

DETERMINATION OF GLUCOSE CONCENTRATION

Blood glucose determination is commonly done in the clinical chemistry laboratories, as there is increasing incidence of diabetes mellitus. It is used to diagnose hyperglycemic conditions like (diabetes mellitus) as well as hypoglycemic situations.

Specimen

Plasma or Serum.

Plasma

Preferred anticoagulant for collecting plasma for glucose estimation is potassium oxalate-sodium fluoride mixture at a ratio of 3:1 and use 3 mg of this mixture/mL blood.

Separate plasma as soon as possible. Never add more than 10 mg of sodium fluoride/mL of blood. Because higher concentrations of sodium fluoride will inhibit color development by glucose oxidase method. Sodium fluoride prevent glycolysis within the cells by inhibiting enolase enzyme of glycolytic pathway. Potassium oxalate prevent clotting by precipitating calcium.

Serum

Plain blood collected in dry containers.

Serum should be separated within 30 minutes of blood collection to avoid getting low glucose values due to glycolysis in blood cells (glucose value decrease at the rate of approximately 7 mg%/hour at room temperature (25–30 °C).

Methods

Enzymatic Methods

- Glucose oxidase method (commonly done)
- Hexokinase method (Reference method).

Reduction Methods

- Folin-Wu method
- Orthotoluidine method.

Enzymatic method of glucose oxidase: Peroxidase method is widely used. Hexokinase method is employed in higher laboratories. Chemical methods are rarely done.

Glucose Oxidase (GOD/POD) Method[1]

Specimen
Plasma or Serum.

Principle
Glucose oxidase reagent contains glucose oxidase (GOD), peroxidase (POD) and mutarotase. Glucose is converted to gluconic acid and hydrogen peroxide by the enzyme glucose oxidase. Hydrogen peroxide then split to form water and nascent oxygen. The nascent oxygen in turn oxidize the chromogen (e.g. 4-aminoantipyrine) to form a pink color. Glucose oxidase has the specificity for β-D glucose. The enzyme mutarotase of this reagent, convert α–D glucose if any in the specimen to β-D glucose. So, this method gives true value of glucose level (Fig. 12.1).

Reagents
- **Reagent kits** based on GOD/POD method is commercially available. Use any one of them and follow the instructions given in the leaflet.
- **Glucose standard (100 mg%).**
- **Working solution:** Provided with the kit. It generally contains glucose oxidase, peroxidase, a dye (here example given is that of 4-aminoantipyrine) and phenol in phosphate buffer of pH 7.

Procedure (Table 12.1)
- Label B, S and T on three separate tubes for Blank, Standard and Test respectively.
- Add 2.0 mL working solution to B, S and T
- Add 0.02 mL standard to 'S' tube. Mix well
- Add 0.02 mL plasma/serum to 'T' tube. Mix well.

Calculation

$$\text{Glucose in mg\%} = \frac{\text{Absorbance of T}}{\text{Absorbance of S}} \times$$

concentration of S in 0.02 mg × $\dfrac{100}{\text{volume of serum taken}}$

$$\text{Glucose in mg\%} = \frac{\text{Absorbance of T}}{\text{Absorbance of S}} \times 0.02 \text{ mg} \times$$

$$\frac{100}{0.02} = \frac{T}{S} \times 100 \text{ mg\%}$$

Advantages of Glucose Oxidase Method Over Reduction Methods
- Glucose oxidase being very specific for β-D glucose, gives the true value of glucose (not including other carbohydrate and non-carbohydrate substances present in the

TABLE 12.1 Procedure of GOD–POD method of blood glucose estimation

	T	S	B
Plasma/Serum	0.02 mL (20 µL)	–	–
Standard	–	0.02 mL (20 µL)	–
Working solution	2 mL	2 mL	2 mL

Incubation: Incubate the assay mixture for 15 minutes at 37°C in an incubator or 30 minutes at room temperature (25°–30°C).

Reading: Measure the absorbance of T and S against B at 505 nm or with green filter. Final color is stable for 2 hours if not exposed to light.

β - D Glucose + O_2 + H_2O —— Glucose oxidase (GOD) ——→ Gluconic acid + H_2O_2

H_2O_2 + 4-aminoantipyrine —— Peroxidase (POD) ——→ Oxidized 4-aminoantipyrine (Red) + H_2O

Figure 12.1 Principle of glucose oxidase (GOD-POD) method

blood). Whereas reduction based methods are nonspecific and act on all reducing substances present in the blood and the results will be falsely higher than true glucose values.
- Glucose oxidase method can be employed to estimate glucose in cerebrospinal fluid (CSF).

Limitations of Glucose Oxidase Method
- Glucose oxidase method cannot be employed to estimate glucose in urine. It is because urine contains substances like uric acid that interfere with peroxidase reaction → low glucose values. (Hexokinase, Glucose dehydrogenase or o-Toluidine methods can be used for this purpose).

Method of Asatoor and King[2] (Table 12.2)

Specimen: Plasma/Serum/Whole blood.

Principle: A protein-free filtrate is heated with alkaline cupric tartrate solution. Glucose present in the specimen reduces cupric ions to cuprous ions which precipitate as insoluble cuprous oxide. The amount of cuprous oxide formed is measured by the reduction of phosphomolybdate to **molybdenum blue** by adding phosphomolybdic acid.

Reagents required (For details see Reagent Preparation):
- Isotonic sodium sulfate–copper sulfate solution (Isotonic diluent)
- Sodium tungstate 10% (10 g/dL)
- Alkaline tartrate solution (Harding's B solution, half strength)
- Phosphomolybdic acid reagent
- Stock standard solution 100 mg% (in isotonic sodium sulfate-copper sulfate solution)
- Working standard (2.5 mg%): Dilute 2.5 mL stock standard to 100 mL with isotonic sodium sulfate-copper sulfate solution.

Procedure
- **Deproteinization of test sample and setting up "Test" (T) tube:** Add 0.1 mL of whole blood or serum or plasma into 3.8 mL of isotonic sodium sulfate–copper sulfate solution taken in a centrifuge tube. Mix well. Add 0.1 mL of 10% sodium tungstate, mix and centrifuge. Thus original blood/serum/plasma is diluted. (Volume of blood/serum/plasma present in 1 mL of diluted sample = 0.1/4 = 0.025 mL).

TABLE 12.2 Procedure of Asatoor and King method of blood glucose estimation in Folin–Wu tubes

	T	S	B
Isotonic diluent	–	–	1 mL
Protein free supernatant	1 mL	–	–
Standard	–	1 mL	–
Alkaline tartrate solution	1 mL	1 mL	1 mL
Mix well. Plug with cotton wool. Keep the tubes in a boiling water bath for **10 minutes**. Then cool the tubes quickly by keeping in water without shaking			
Phosphomolybdic acid solution	3 mL	3 mL	3 mL
Distilled water	3 mL	3 mL	3 mL
Mix thoroughly by inverting the tubes			
Reading: Read at 680 nm or using a red filter			

Transfer **1.0 mL of protein free supernatant** into Folin-wu tube. Mark this tube as 'T' for test sample.
- **Setting up blank (B) and standard (S) tubes**
  Blank (B) tube—Pipette **1.0 mL** isotonic sodium sulfate-copper sulfate solution into 'B' tube.
  Standard (S) tube—Pipette **1.0 mL** working standard into 'S' tube.
- To all the three tubes (B, S and T) add 1 mL alkaline tartrate solution. Mix well. Plug with cotton wool. Keep the tubes in a boiling water bath for **10 minutes**. Then cool the tubes quickly by keeping in water without shaking. Add 3.0 mL of phosphomolybdic acid solution and 3.0 mL distilled water to each tube (precipitated cuprous oxide dissolves). Mix thoroughly by inverting the tubes.
- **Reading:** Read at 680 nm or using a red filter.

Calculation

$$\text{Glucose in mg\%} = \frac{\text{Absorbance of T}}{\text{Absorbance of S}} \times \text{concentration of S in 1.0 mL} \times \frac{100}{\text{volume of serum taken}}$$

$$\text{Glucose in mg\%} = \frac{\text{Absorbance of T}}{\text{Absorbance of S}} \times 0.025 \text{ mg} \times \frac{100}{0.025} = \frac{T}{S} \times 100 \text{ mg\%}$$

Reference Interval of Fasting Glucose in Adults

Serum/Plasma: 74–100 mg% (4.1–5.6 mmol/L).

Whole blood: 65–95 mg% (3.6–5.3 mmol/L).

(To convert mg% into mmol/L multiply by conversion factor 0.0555).

Points to Ponder
- For the diagnosis of diabetes mellitus, the reference intervals are not used. Diagnostic criteria in terms of blood glucose laid down by American Diabetes Association (ADA)[3] is used for this purpose.
- Whole blood is not a preferred specimen for the blood glucose determination. Glucose being water soluble, it associates with water. Water content in plasma is 93% and in RBC 73%. Hence, whole blood if used, the glucose values obtained vary with packed cell volume of the blood.

GLUCOSE TOLERANCE TEST

The oral glucose tolerance test (OGTT) evaluates clearance of glucose from the blood circulation after oral glucose loading under standard conditions. The Committee on statistics of the American Diabetes Association (ADA) has standardized the test.

Standard Conditions

- A minimum carbohydrate intake of 150 g/day for 3 days should be taken before the test (Otherwise carbohydrate tolerance will be lowered giving falsely high results)
- The subject should be on 8–16 hours fast before testing
- The person should be on routine activities and not bed ridden (because inactivity decreases glucose tolerance)
- The person should be peaceful without any emotional stress
- Should avoid exercise
- Should be free of illness since illness will reduce glucose tolerance. Abnormalities involving thyroxin, growth hormone, cortisol and catecholamines will interfere with the test
- Drugs like oral contraceptives, hypoglycemic agents (sulfonylureas, insulin), diuretics, salicylates and other agents like tobacco and caffeine will interfere with the test. These must be stopped prior to the test
- Test should begin between 7 AM and 9 AM

- Age must be considered during interpretation of the test. Adjustments for age should be done
- Glucose load consists of glucose only. Other forms of carbohydrates are not recommended.

During the test: Patient should sit quietly. Should not smoke (smoking elevates blood glucose).

Indications for Oral Glucose Tolerance Test (OGTT)

Not recommended as a routine test. There are specific indications for doing OGTT.
- Diagnosis of gestational diabetes mellitus
- Diagnosis of impaired glucose tolerance
- Diagnosis of renal glucosuria
- Population studies for collecting epidemiological data.

Procedure

- Collect fasting blood and urine samples
- **Dose:** 75 g of anhydrous glucose for adults, 100 g for pregnant women and 1.75 g/kg body weight for children, may be dissolved in 300 mL of water and it may be taken within 15 minutes time. Can be flavored with lime or orange juice.
- Collect blood and urine samples at ½ hour intervals after the intake of glucose load for up to 2 hours.
- Test urine for glucose by either Benedict's qualitative test or glucose oxidase based strip test (ideal) and estimate glucose in all plasma samples.

Interpretation of GTT

Plasma Glucose and Urine Glucose in Health (Table 12.3)

Urine glucose: Absent in all the samples.

Fasting blood glucose: <100 mg% (<5.5 mmol/L).

2 hour postprandial blood glucose: <140 mg% (<7.8 mmol/L).

Different Types of OGTT Curves (Figs 12.2 to 12.6)

Conditions Causing Abnormal Glucose Tolerance

- **Diabetes mellitus** (Fig. 12.5)
- **Impaired fasting glucose:** Follow-up required because they are prone for developing diabetes mellitus in future.
- **Impaired glucose tolerance:** Follow-up required because they are prone for developing diabetes mellitus in future.
- **Increased glucose tolerance (flat OGTT curve) (Fig. 12.6):** Normally after an glucose load, blood glucose values rise to peak at 60 minutes and then fall to near fasting levels

TABLE 12.3 Plasma glucose levels in OGTT in normal, diabetes mellitus, IGT and IFG

	Normal	Diabetes mellitus	Impaired glucose tolerance (IGT)	Impaired fasting glucose (IFG)
Fasting	<99 mg% (<5.5 mmol/L)	>126 mg% (7 mmol/L)	Between 100 and 125 mg% (6.1 mmol/L and 7 mmol/L)	*Between 100 and 125 mg% (6.1 mmol/L and 7 mmol/L)
1 Hour	<160 mg% (9 mmol/L)	Not set	Not set	
2 Hours	<140 mg% (7.8 mmol/L)	>200 mg% (11.1 mmol/L)	Between 140 and 199 mg% (7.8 mmol/L and 11.1 mmol/L)	

* ADA recommendation in 2003: Lowered the lower limit of Impaired Fasting Glucose from 110 to 100 mg%.

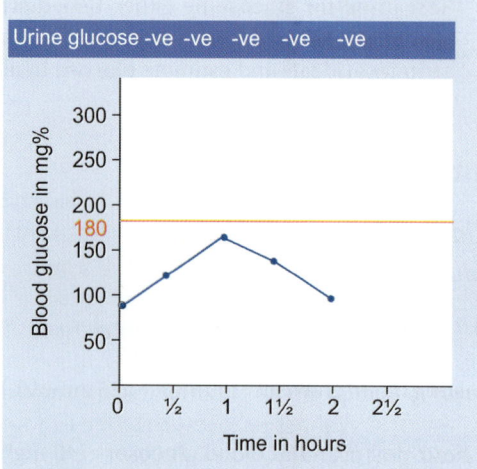

Figure 12.2 Oral glucose tolerance test curve: Normal

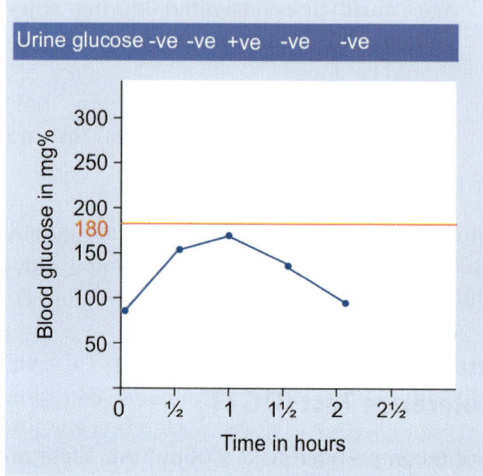

Figure 12.4 Oral glucose tolerance test curve: Renal glucosuria

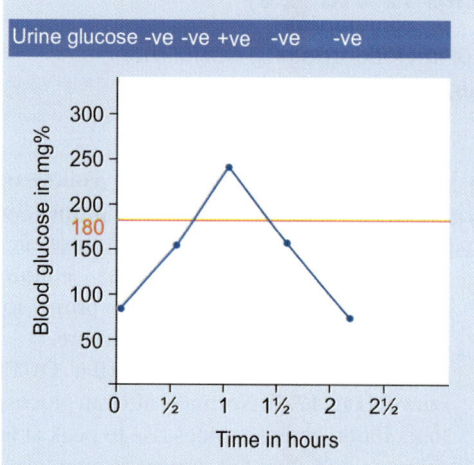

Figure 12.3 Oral glucose tolerance test curve: Alimentary glucosuria

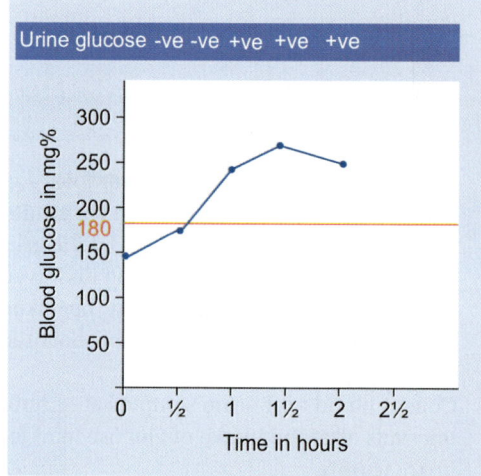

Figure 12.5 Oral glucose tolerance test curve: Diabetes mellitus

at 2 hours. But in some cases blood glucose show only a minimal rise at 60 minutes. This type of OGTT curve obtained in malabsorption, hypopituitrism, Addison's disease and hypothyroidism.

- **Alimentary glucosuria or lag curve (Fig. 12.3):** Exaggerated rise in glucose value after ingestion of glucose load within 1–1 ½ hours, even crossing the renal threshold for glucose leading to excretion of glucose in urine. But by 2 hours blood glucose level goes down to normal levels or even to hypoglycemic levels. This is due to conditions leading to rapid emptying

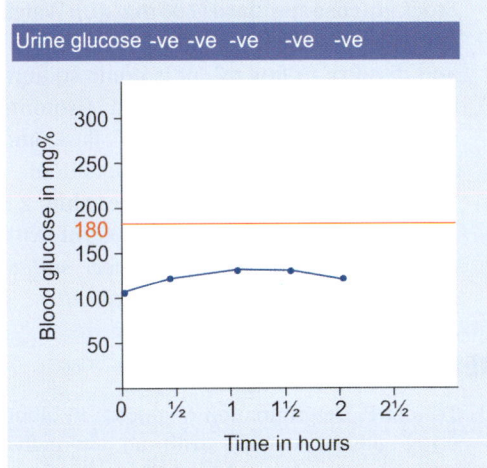

Figure 12.6 Oral glucose tolerance test curve: Flat curve

of stomach causing increased rate of glucose absorption from the gut, e.g: hyperthyroidism, partial gastrectomy.
- **Renal glucosuria (Fig. 12.4):** It occurs in persons with lowered renal threshold for glucose. Normal renal threshold for glucose is 180 mg%. In some individuals it is lowered so that at normal blood glucose levels, glucose is excreted in urine. Most often it is discovered, by chance. The condition will not cause any harm to the person. It is frequently encountered in the third trimester of pregnancy.

QUESTIONS

1. What are the methods used to estimate glucose in the blood. Which is the best method among them? Why?
2. Give the normal values of fasting and 2 hr postprandial blood glucose?
3. Name the chemicals to be added into the container meant for blood collection for glucose estimation.
4. Give the principle of glucose oxidase method of blood glucose estimation.
5. What is the ADA criteria for the diagnosis of diabetes mellitus?
6. Classify diabetes mellitus.
7. What is glucose tolerance test? Give three indications for doing GTT.
8. What is impaired glucose tolerance test? How will you confirm it?
9. What are the instructions to be given to a patient advised for an OGTT?
10. Give briefly the procedure of OGTT.
11. What do you mean by renal glucosuria?
12. What is meant by alimentary glucosuria?
13. When will you get a flat curve on GTT? Mention the conditions causing it.
14. Diagnostic criteria of diabetes mellitus.

REAGENT PREPARATION

- **Isotonic sodium sulfate–copper sulfate solution (Isotonic diluent):** Dissolve 30 g of hydrated sodium sulfate (Na_2SO_4 10 H_2O) or 13.23 g of anhydrous sodium sulfate and 6 g of copper sulfate ($CuSO_4$ 5 H_2O) in 1 L of distilled water.
- **Sodium tungstate (10 g/dL):** Dissolve 10 g of sodium tungstate (Na_2WO_4 2 H_2O) in about 80 mL of distilled water and then make up to 100 mL distilled water.
- **Alkaline tartrate solution (Harding's B solution, half strength):**
  Contains the following in 1 L of distilled water
  a. Sodium bicarbonate—25 g
  b. Sodium carbonate (anhydrous)—20 g
  c. Potassium oxalate—18 g
  d. Sodium potassium tartrate—12 g

  Dissolve **25 g of sodium bicarbonate** in 600 mL of distilled water in a one liter flask at room temperature. Add **20 g sodium carbonate (anhydrous)**. Dissolve it without heating. Weigh **18 g of potassium oxalate.** Dissolve potassium oxalate by adding small

amounts in warm distilled water in a beaker, transferring each time into the bicarbonate-carbonate mixture. Dissolve **12 g of sodium potassium tartrate** separately in a little amount of distilled water in a beaker and add this to the former mixture in one liter flask and make up to 1000 mL with distilled water and mix well. It is stable for 1 year at 25–30°C.

- **Phosphomolybdic acid reagent:** Dissolve 35 g molybdic acid and 5 g sodium tungstate in 200 mL of 10% sodium hydroxide and 200 mL of distilled water and boil for 45 minutes to remove ammonia present in the molybdic acid. Add a few glass beads to prevent bumping and spilling during boiling. Cool and transfer to a 500 mL flask and dilute it up to 350 mL, with water. Then add 125 mL phosphoric acid (S.G.1.75) and make up to 500 mL with water. It is stable for 1 year at 25–30°C.

- **Stock glucose standard (100 mg/dL):** Weigh 100 mg dry anhydrous glucose (dextrose) and dissolve in few mL of isotonic sodium sulfate-copper sulfate solution (isotonic diluent) in a 100 mL standard flask and make up to 100 mL with the same diluent.
- **Working standard (2.5 mg/dL):** Dilute 2.5 mL stock glucose standard to 100 mL with isotonic sodium sulfate–copper sulfate solution in a 100 mL standard flask.

REFERENCES

1. Trinder P. Determination of glucose in blood using glucose oxidase with an alternative oxygen acceptor. Ann Clin Biochem. 1967;6:24-7.
2. Asatoor AM, King EJ. Simplified colorimetric blood sugar method. Biochem J. 1954;56:xliv.
3. Report of the expert committee (ADA) on the diagnosis and classification of diabetes mellitus. Diabetes Care. 1997;20:1183-201.

CHAPTER 13

Determination of Urea

UREA

Urea is the end product of protein catabolism. Deamination of amino acids release ammonia which is detoxified in the liver to form urea (Fig. 13.1) and more than 90% of urea produced is excreted in urine and the rest through gastrointestinal tract and skin. It is filtered freely at the glomeruli and neither actively reabsorbed nor secreted by the tubules. However, 40–70% of the filtered urea reenters plasma by passively diffusing out of the renal tubule into the interstitium. Urine flow rate also influences this back diffusion. Higher the urine flow lesser the back diffusion. Hence, urea clearance under-estimate glomerular filtration rate (GFR). Besides urea level is dependent on diet and hepatic synthesis of urea. If the diet is rich in protein, more urea will be formed and excreted.

Proteins
↓ Protein breakdown
Amino acids
↓ Transdeamination
Ammonia
↓ Urea cycle in liver
Urea

Figure 13.1 Formation of urea

Methods

Enzymatic Methods[1]
- Using urease enzyme combined with Berthelot reaction
- Using urease enzyme combined with glutamate dehydrogenase.

Chemical Methods
Fearon method.[2]

Electrochemical Approach
Using conductimetric method.[3]

Potentiometry
Using urease enzyme and ammonium selective electrode.[4]

Fearon Method[2] using Diacetyl Monoxime

Specimen

Serum or anticoagulated whole blood or plasma.

Method Used

Fearon method using diacetyl monoxime—Thiosemicarbazide.[2]

Principle

It is based on **Fearon reaction** in which diacetyl condenses with urea to give a complex diazine.[1] Diacetyl being unstable, it is provided by diacetyl monoxime, which on heating in acid medium decomposes into diacetyl and hydroxylamine. Thiosemicarbazide and ferric ions added to the reaction mixture, to enhance and stabilize the color (Fig. 13.2).

Reagents Required

(For details see Reagent Preparation)
- **Trichloroacetic acid (3 g/dL)** used as protein precipitant
- **Urea working standard (30 mg%)**
- **Acid reagent**
- **Color reagent:** Contains acid reagent, diacetyl monoxime and thiosemicarbazide

Procedure

To prepare protein free filtrate: Add 0.2 mL blood, serum or plasma to 1.8 mL 3% trichloroacetic acid in a centrifuge tube. Mix well and keep for 5 minutes and then centrifuge. Use the supernatant as test solution in the procedure.

To prepare standard: Add 0.2 mL of working standard to 1.8 mL trichloroacetic acid (3 g/dL). Pipette from this for the experiment.

Take three test tubes and label T (test), S (standard) and B (blank) and proceed as given in Table 13.1.

TABLE 13.1 Procedure of urea estimation using diacetyl monoxime

	T (mL)	S (mL)	B (mL)
Protein free supernatant	0.2	-	-
Standard urea solution	-	0.2	-
Color reagent	5.0	5.0	5.2

Keep these three tubes in a boiling water bath for 20 minutes. Cool the tubes to room temperature. Take the reading in a photoelectric colorimeter using green filter in a colorimeter or at 535 nm in a spectrophotometer

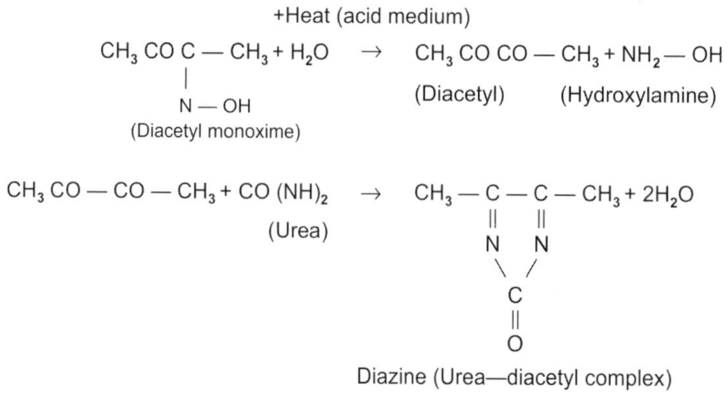

$$CH_3\text{-}CO\text{-}C(NOH)\text{-}CH_3 + H_2O \xrightarrow{+\text{Heat (acid medium)}} CH_3\text{-}CO\text{-}CO\text{-}CH_3 + NH_2\text{-}OH$$

(Diacetyl monoxime) → (Diacetyl) (Hydroxylamine)

$$CH_3\text{-}CO\text{-}CO\text{-}CH_3 + CO(NH_2)_2 \rightarrow \text{Diazine} + 2H_2O$$

(Urea) → Diazine (Urea—diacetyl complex)

Figure 13.2 Principle of diacetyl monoxime—thiosemicarbazide method of urea estimation

Calculation

Concentration of urea in 0.2 mL of working standard = 0.06 mg

0.06 mg/0.2 mL diluted to 2.0 mL in the procedure

∴ concentration of standard after dilution = 0.006 mg/0.2 mL

Concentration of urea in 100 mL blood (mg %)

$$= \frac{\text{OD of T} - \text{OD of B}}{\text{OD of S} - \text{OD of B}} \times \text{Concentration of standard} \times 100/\text{vol of serum taken}$$

$= T/S \times 0.006 \times 100/0.02$ mg%

$= T/S \times 30$ mg%

Reference Interval

Blood urea concentration in healthy individuals is between 15–40 mg%. Values near upper limits of reference range are seen with high protein intake.

Interpretation

The concentration of urea in the whole blood is slightly less than that in plasma or serum. Urea being soluble, is distributed in intracellular and extracellular water. Since there is less water inside the blood cells, the concentration of urea in the whole blood is lower. Measurement of urea alone is less useful in diagnosing kidney diseases because its blood level is influenced by dietary proteins and hepatic function. But its diagnostic value improves along with serum creatinine values.

BUN (Blood Urea Nitrogen)

Nowadays urea is expressed in terms of BUN (blood urea nitrogen)

Molecular weight of urea (NH_2–CO–NH_2) = 60

Hence, 60 g of urea contains 28 g of nitrogen

To convert urea in mg% into BUN multiply by 28/60, i.e. 0.467

To convert BUN into urea in mg% multiply by 60/28, i.e. 2.14

Utility of Urea Measurement in Clinical Medicine

- Assessment of kidney function: More useful when urea values are interpreted along with serum creatinine values.
- To differentiate between prerenal and postrenal uremia.

For this purpose a ratio is used → **Serum urea nitrogen mg%/Serum creatinine mg%.**

For a healthy individual on a normal diet, this **ratio ranges from 12 to 20.**

Low ratios of serum urea nitrogen mg%/serum creatinine mg% seen in:
- Acute tubular necrosis
- Low protein intake
- Starvation
- Severe liver disease → ↓ urea synthesis.

High ratios of serum urea nitrogen mg%/serum creatinine mg% with normal serum creatinine seen in:
- ↑ Catabolic states or ↑ Tissue breakdown
- Prerenal causes of uremia
- High protein intake
- Gastrointestinal tract hemorrhage.

High ratios of serum urea nitrogen mg%/serum creatinine mg% with ↑ serum creatinine seen in:
- Postrenal causes of uremia
- Prerenal causes of uremia coexisting with postrenal causes.

Alterations in Plasma/Urea Concentration

Causes of low levels of blood urea:
- Low protein intake
- Conditions leading to hemodilution
- Severe liver disease causing impaired urea cycle.

Causes of high levels of blood urea: It is referred to as Uremia. It is seen with disorders of kidney in which the GFR is reduced (Fig. 13.3).

QUESTIONS

1. What is the method used to estimate urea in the blood? Give its principle.
2. Give the normal values of blood urea.
3. Why do the whole blood urea values are lower than plasma or serum urea levels?
4. What is uremia? What are the different causes of uremia?
5. Is blood urea level a critical diagnostic marker of kidney disease? If not, give reason.
6. What are the prerenal causes of uremia? Explain the mechanism of the causation uremia in these conditions.
7. Mention three renal causes of uremia.
8. Mention three postrenal causes of uremia. Explain the mechanism.
9. Name the factors affecting urea level in the blood in a normal person.
10. Urea clearance under estimate GFR. Explain.

REAGENT PREPARATION

- **Trichloroacetic acid (TCA) (3 g/dL):** Weigh 3 g of TCA and dissolve in a few mL of distilled water in glass cylinder and make up to 100 mL with distilled water. It is stable up to 1 year.
- **Orthophosphoric acid (specific gravity 1.750, purity 85–90%):** It is corrosive and should be handled with care.
- **Ferric chloride (5 g/dL):** Dissolve 5 g of anhydrous ferric chloride in 80 mL of distilled water. Add 1 mL of concentrated sulfuric acid make up to 100 mL with distilled water.
- **Concentrated sulfuric acid (specific gravity 1.840, purity 98%):** It is corrosive and should be handled with care.
- **Diacetyl monoxime (2.5 g/dL):** Dissolve 2.5 g diacetyl monoxime in distilled water and make up to 100 mL with distilled water in a volumetric flask. If stored in amber colored bottles at 25–30° C, it is stable up to 6 months.
- **Thiosemicarbazide (0.25 g/dL):** Weigh out 0.25 g of thiosemicarbazide and dissolve in a few mL of distilled water and make up to 100 mL with distilled water in a volumetric flask. If stored at 25–30° C, it is stable up to 6 months.

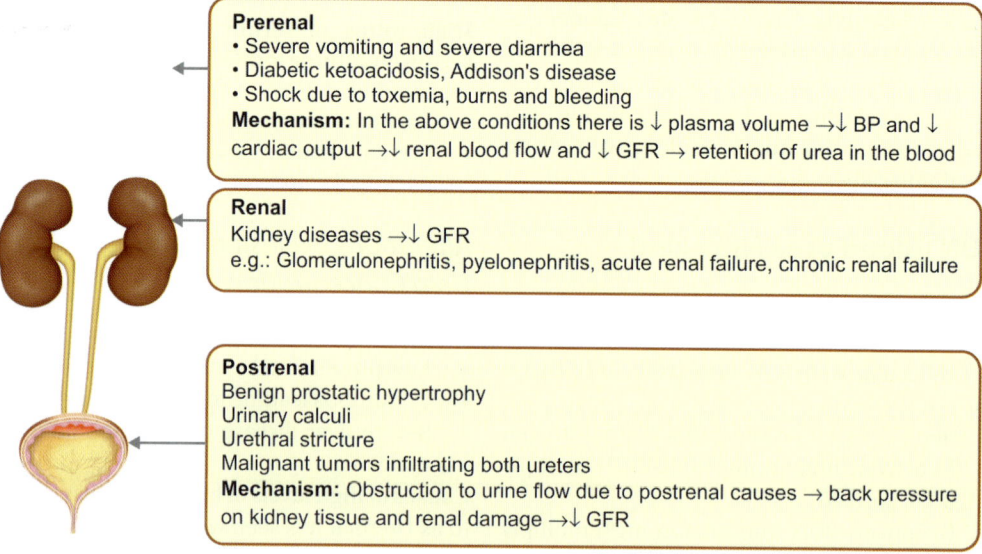

Prerenal
- Severe vomiting and severe diarrhea
- Diabetic ketoacidosis, Addison's disease
- Shock due to toxemia, burns and bleeding

Mechanism: In the above conditions there is ↓ plasma volume →↓ BP and ↓ cardiac output →↓ renal blood flow and ↓ GFR → retention of urea in the blood

Renal
Kidney diseases →↓ GFR
e.g.: Glomerulonephritis, pyelonephritis, acute renal failure, chronic renal failure

Postrenal
Benign prostatic hypertrophy
Urinary calculi
Urethral stricture
Malignant tumors infiltrating both ureters

Mechanism: Obstruction to urine flow due to postrenal causes → back pressure on kidney tissue and renal damage →↓ GFR

Figure 13.3 Causes of uremia

Calculation

Concentration of urea in 0.2 mL of working standard = 0.06 mg

0.06 mg/0.2 mL diluted to 2.0 mL in the procedure

∴ concentration of standard after dilution = 0.006 mg/0.2 mL

Concentration of urea in 100 mL blood (mg %)

$$= \frac{\text{OD of T} - \text{OD of B}}{\text{OD of S} - \text{OD of B}} \times \text{Concentration of standard} \times 100/\text{vol of serum taken}$$

$= T/S \times 0.006 \times 100/0.02$ mg%

$= T/S \times 30$ mg%

Reference Interval

Blood urea concentration in healthy individuals is between 15–40 mg%. Values near upper limits of reference range are seen with high protein intake.

Interpretation

The concentration of urea in the whole blood is slightly less than that in plasma or serum. Urea being soluble, is distributed in intracellular and extracellular water. Since there is less water inside the blood cells, the concentration of urea in the whole blood is lower. Measurement of urea alone is less useful in diagnosing kidney diseases because its blood level is influenced by dietary proteins and hepatic function. But its diagnostic value improves along with serum creatinine values.

BUN (Blood Urea Nitrogen)

Nowadays urea is expressed in terms of BUN (blood urea nitrogen)

Molecular weight of urea (NH_2–CO–NH_2) = 60

Hence, 60 g of urea contains 28 g of nitrogen

To convert urea in mg% into BUN multiply by 28/60, i.e. 0.467

To convert BUN into urea in mg% multiply by 60/28, i.e. 2.14

Utility of Urea Measurement in Clinical Medicine

- Assessment of kidney function: More useful when urea values are interpreted along with serum creatinine values.
- To differentiate between prerenal and postrenal uremia.

For this purpose a ratio is used → **Serum urea nitrogen mg%/Serum creatinine mg%.**

For a healthy individual on a normal diet, this **ratio ranges from 12 to 20.**

Low ratios of serum urea nitrogen mg%/serum creatinine mg% seen in:
- Acute tubular necrosis
- Low protein intake
- Starvation
- Severe liver disease → ↓ urea synthesis.

High ratios of serum urea nitrogen mg%/serum creatinine mg% with normal serum creatinine seen in:
- ↑ Catabolic states or ↑ Tissue breakdown
- Prerenal causes of uremia
- High protein intake
- Gastrointestinal tract hemorrhage.

High ratios of serum urea nitrogen mg%/serum creatinine mg% with ↑ serum creatinine seen in:
- Postrenal causes of uremia
- Prerenal causes of uremia coexisting with postrenal causes.

Alterations in Plasma/Urea Concentration

Causes of low levels of blood urea:
- Low protein intake
- Conditions leading to hemodilution
- Severe liver disease causing impaired urea cycle.

Causes of high levels of blood urea: It is referred to as Uremia. It is seen with disorders of kidney in which the GFR is reduced (Fig. 13.3).

QUESTIONS

1. What is the method used to estimate urea in the blood? Give its principle.
2. Give the normal values of blood urea.
3. Why do the whole blood urea values are lower than plasma or serum urea levels?
4. What is uremia? What are the different causes of uremia?
5. Is blood urea level a critical diagnostic marker of kidney disease? If not, give reason.
6. What are the prerenal causes of uremia? Explain the mechanism of the causation uremia in these conditions.
7. Mention three renal causes of uremia.
8. Mention three postrenal causes of uremia. Explain the mechanism.
9. Name the factors affecting urea level in the blood in a normal person.
10. Urea clearance under estimate GFR. Explain.

REAGENT PREPARATION

- **Trichloroacetic acid (TCA) (3 g/dL):** Weigh 3 g of TCA and dissolve in a few mL of distilled water in glass cylinder and make up to 100 mL with distilled water. It is stable up to 1 year.
- **Orthophosphoric acid (specific gravity 1.750, purity 85–90%):** It is corrosive and should be handled with care.
- **Ferric chloride (5 g/dL):** Dissolve 5 g of anhydrous ferric chloride in 80 mL of distilled water. Add 1 mL of concentrated sulfuric acid make up to 100 mL with distilled water.
- **Concentrated sulfuric acid (specific gravity 1.840, purity 98%):** It is corrosive and should be handled with care.
- **Diacetyl monoxime (2.5 g/dL):** Dissolve 2.5 g diacetyl monoxime in distilled water and make up to 100 mL with distilled water in a volumetric flask. If stored in amber colored bottles at 25–30° C, it is stable up to 6 months.
- **Thiosemicarbazide (0.25 g/dL):** Weigh out 0.25 g of thiosemicarbazide and dissolve in a few mL of distilled water and make up to 100 mL with distilled water in a volumetric flask. If stored at 25–30° C, it is stable up to 6 months.

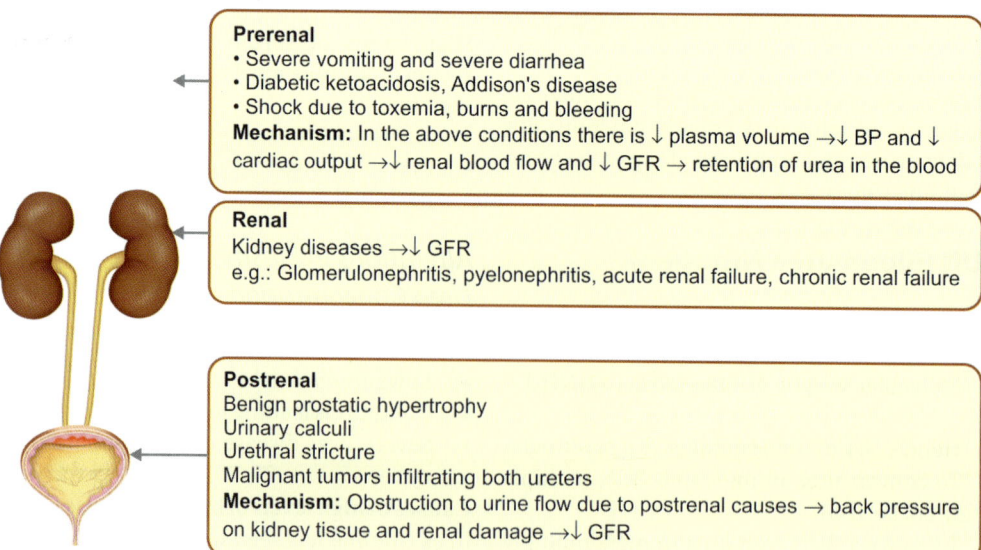

Figure 13.3 Causes of uremia

- **Benzoic acid solution (2.5 g/L):** Add about 2.5 g of benzoic acid to 1 L of hot distilled water.
- **Urea standard: Stock standard (1 g/dL):** Weigh accurately 1 g of pure urea (AR) and dissolve in a few mL of benzoic acid solution and make up to 100 mL with benzoic acid solution in a standard flask. Store at 25–30° C → stable up to 6 months.
- **Working standard (30 mg/dL):** Dilute 3 mL of stock standard to 100 mL in a standard flask with benzoic acid solution.
- **Acid reagent:** Take 300 mL of distilled water in a volumetric flask and add with caution 24 mL of concentrated sulfuric acid and 60 mL of orthophosphoric acid and 3 mL of ferric chloride solution (5 g/dL). It is stable for 24 hours only.
- **Color reagent:** Since this reagent is stable for 3 hours only, make only required volume. Mix 75 mL acid reagent, 50 mL distilled water, 2.5 mL diacetyl monoxime (2.5 g/dL) and 0.6 mL of thiosemicarbazide (0.25 g/dL).

REFERENCES

1. Sampson EJ, Baird MA. Chemical inhibition used in a kinetic urease/glutamate dehydrogenase method for urea in serum. Clin Chem. 1979;25:1721-9.
2. Veniamin MP, Vakirtzi – Lemonias C. Chemical basis of the carbamido diacetyl micromethod for estimation of urea, citrulline and carbamyl derivatives. Clin Chem. 1970;16:3-6.
3. Eckfeldt J, Levine AS, Greiner C, et al. Urinary urea: Are currently available methods adequate for revival of almost abandoned test? Clin Chem. 1982;28:1500-2.
4. Rovida E, Mosaca A, Dossi G, et al. Serum and whole blood urea determination by the use of microprocessor controlled differential pH analyzer. Eur J Clin Chem Clin Biochem. 1981;19:820.

CHAPTER 14

Determination of Creatinine

CREATININE

Creatine, methyl guanidinoacetic acid is synthesized in the liver and kidney and carried by the blood to muscular tissues and brain and converted to creatine phosphate. Energy needed for muscular contraction is provided by ATP break down to form ADP. The ATP is regenerated from ADP by the action of creatine kinase. This regeneration of ATP by the hydrolysis of creatine phosphate is called Lohmann's reaction (Fig. 14.1). During this process, creatine phosphate is converted to creatine. Creatine in turn converted by spontaneous dehydration into creatinine. About 2% of the total creatine is converted to creatinine per day so that the rate of creatinine formation is constant in an individual as it is related to muscle mass.

Creatine is filtered at the glomerulus and reabsorbed by the tubules. So that at low plasma concentrations, as in normal individuals, no creatine appears in urine. **Creatinine** is also filtered at the glomerulus. It is reabsorbed at PCT in very small amounts and secreted in the tubules to a minor degree and the rest is excreted in urine.

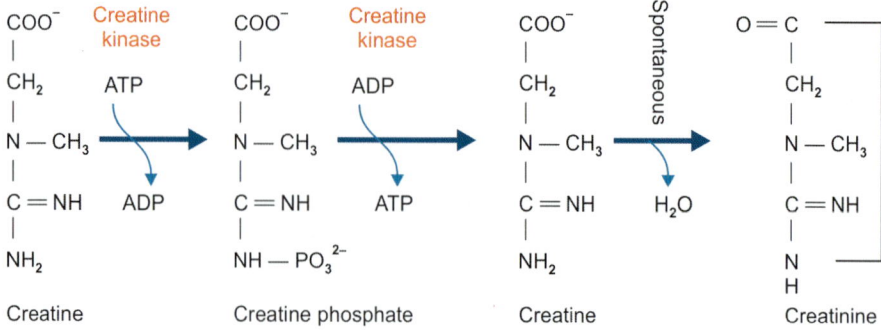

Figure 14.1 Creatine phosphate and Lohmann's reaction

Methods[1,2]

- **Chemical methods**
  - Jaffe's reaction based method.
- **Enzymatic methods**
  - Creatininase based method
  - Creatininase and creatinase based method
  - Creatinine deaminase based method.

Jaffe's Reaction Based Method[1]

Specimen

Serum or plasma. Whole blood should not be used because blood cells contain substances that interfere with Jaffe's reaction.

Principle

In Jaffe's reaction an orange-red colored complex is formed when creatinine is allowed to react with an alkaline picrate solution. The absorbance of this complex is measured with green filter in a colorimeter or at 505 nm in a spectrophotometer.

Reagents required (For details Reagent Preparation)

All reagents should be of analytical reagent grade (AR).

- Sodium hydroxide solution (2.8 g/dL or 0.7 mol/L)
- Picric acid solution (0.04 mol/L)
- Acid tungstate reagent
- Stock creatinine standard (100 mg/dL)
- Working creatinine standard (1 mg/dL).

Procedure

Preparation of protein free supernatant: Add **0.5 mL of plasma or serum** to 4 mL of acid tungstate solution taken in a centrifuge tube. Mix well and centrifuge at 3000 rpm for 10 minutes to get a clear supernatant.

Set tubes T, B and S in the manner shown in Figure 14.2.

- Transfer 3 mL of the supernatant (contains 0.375 mL serum/plasma) into the test tube labeled **'T'**
- Add **3.0 mL of distilled water** into a test tube labeled **'B'**
- Add **0.5 mL of working standard (1 mg/dL)** + 2.5 mL distilled water into test tube labeled **'S'**
- Then add **1.0 mL picric acid** solution and mix well
- Add **1.0 mL sodium hydroxide** (0.7 mol/L) and shake well
- Keep all the tubes at room temperature for **15 minutes.**

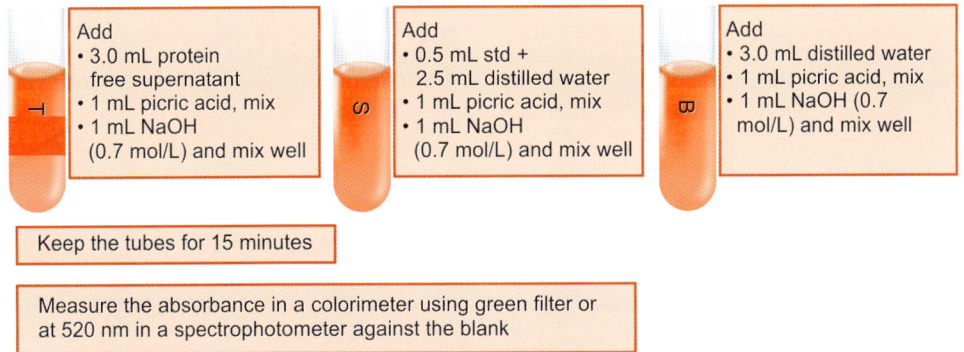

Figure 14.2 Procedure of creatinine assay by Jaffe's reaction based method

- Adjust the reading in the colorimeter to zero with the blank, using green filter in the colorimeter or at **520 nm** in a spectrophotometer. Read the absorbance of test and standard against the blank.

Calculation

Concentration of creatinine in 100 mL blood (mg%) =
OD of T/OD of S × concentration of standard in mg% × 100/volume of sample
   = OD of T/OD of S × 0.005 mg% × 100/0.375
   = OD of T/OD of S × 1.3 mg%.

Reference Interval

Reference range serum—0.7 to 1.4 mg%.

Interpretation

Determination of creatinine levels in the serum or plasma is minimally affected by dietary pattern and state of hydration of the body. More-over, it is mainly excreted through urine.

The plasma/serum creatinine **increases** with renal diseases—nephritis, nephrotic syndrome, acute and chronic renal failure and other types of renal insufficiency caused by drugs and toxins.

Limitations of Jaffe Based Creatinine Assay

- Jaffe's reaction is not specific for creatinine. Many compounds in blood produce Jaffe like chromogens, e.g. Proteins, glucose, ascorbic acid, acetone, acetoacetate, pyruvate, cephalosporins.
- Meat-based diet will increase the creatinine value. Hence, creatinine estimation on fasting serum/plasma sample is ideal.

Creatinine Clearance

It is a test done to assess the renal function. The affection of glomerular filtration membrane is the most important cause of reduced renal function. When glomerulus is involved in disease processes, glomerular filtration will be reduced and consequently excretion of waste products will be affected. So assessment of glomerular filtration is a good tool to recognize any altered renal function.

Renal clearance of a substance is the volume of plasma from which the substance is cleared completely by the kidneys per unit of time.

Urine is formed by filtration at the glomeruli. Except the cellular portion, proteins and lipids and all other constituents of blood pass to the ultrafiltrate at the level of glomerulus. Subsequent passage through the tubule allows reabsorption of most of the water, glucose, sodium, calcium, phosphorous and chloride into the bloodstream to maintain their normal levels in the blood. Mainly the nitrogenous waste products are excreted in urine.

At the level of kidneys, the possible mechanisms dealt by different types of constituents in the blood are given below:

- Filtration at the glomeruli and reabsorption by the tubules
- Filtration at the glomeruli and secretion by the tubules
- Filtration at the glomeruli only and no change (neither reabsorption nor secretion) at the level of tubules—whatever filtered is passed into the urine.

Creatinine belongs to the 3rd group. It is produced within the body and released into the body fluids. It is filtered at the glomerulus and reabsorbed at PCT in small amounts and secreted in the tubules to a minor degree.

To measure creatinine clearance a **timed urine** and **blood specimens** are required.

The volume of urine is measured in terms of mL and creatinine is measured in mg/dL in both urine and serum/plasma specimens.

The creatinine clearance is then calculated by using the formula,

Creatinine clearance (mL/min)
= Creatinine in urine (U) (mg/mL) × V (mL/min)/Creatinine in serum (P)(mg/mL) = UV/P

Reference Interval for Creatinine Clearance

Men: 94–140 mL/min/1.73 m^2
Women: 72–110 mL/min/1.73 m^2

Creatinine clearance can also be calculated (estimated GFR/eGFR) by using **Cockcroft and Gault formula** if serum creatinine concentration, age and weight of an individual are known.

Creatinine clearance (mL/min/1.73 m^2) = (140–age in years) × 2.12 × wt in kg × K ÷ serum creatinine (mg/dL) × Body surface area (m^2)

K = 0.85 for women and 1.0 for men.

Points to Ponder

Since creatinine is produced as a result of muscle contraction, its concentration in the body fluids is related to muscle mass. Hence, men have higher creatinine levels than women.

Urine Creatinine Determination

Since the creatinine level in the body fluids is minimally influenced by the diet, and is excreted at constant rate in an individual, its measurement in urine is useful to check the reliability of 24-hour urine collections.

QUESTIONS

1. What is the method used to estimate creatinine in the blood? Give its principle.
2. Give the reference range of serum creatinine.
3. What is the reason for low creatinine concentrations in females than in males?
4. What is the source of creatinine in the body?
5. Why creatinine is preferred than blood urea in assessing renal function?
6. What do you mean by creatinine clearance? How will you use it?
7. What is the normal creatinine clearance?
8. Give reference interval of creatinine clearance.
9. What are the different uses of estimating creatinine in urine?
10. What is Lohmann's reaction?

REAGENT PREPARATION

- **Sodium hydroxide solution (2.8 g/dL or 0.7 mol/L):** Take 28 g of sodium hydroxide in a measuring cylinder and dissolve in a few mL of distilled water and make up to 1 L with distilled water. Store in a stoppered polyethylene bottle. It is stable up to 1 year at 25–30°C.
- **Picric acid solution (0.04 mol/L):** Dissolve 9.16 g of hydrated picric acid or 8.25 g of anhydrous picric acid in a little of distilled water in a measuring jar and make up to 1 L. Transfer to an amber colored bottle and it is stable for 1 year at room temperature.
  (**Precaution**: Dry picric acid is explosive on percussion. Do not use ground glass stoppers for the containers in which picric acid is kept. Discharging picric acid waste through copper pipes, will cause formation of copper picrate accumulation of which may cause explosion.
- **Sodium tungstate–5%:** Dissolve 5 g of sodium tungstate dihydrate (Na$_2$WO$_4$ 2H$_2$O) in a few mL of distilled water in a measuring cylinder and make up to 100 mL distilled water.

- **Stock creatinine standard (100 mg/dL):** Dissolve 100 mg pure anhydrous creatinine in a few mL of hydrochloric acid (0.1 M) in a volumetric flask and make up to 100 mL with the HCl (0.1 M). It is stable for 6 months at 2–8°C.
- **Working creatinine standard:** Dilute 1 mL of stock creatinine standard to 100 mL in a standard flask using HCl (0.1 M/L). This solution contains 0.01 mg of creatinine/mL (1 mg/100 mL).

REFERENCES

1. DiGiorgio J. Nonprotein nitrogenous constituents. In: Henry RJ, Cannon DC, Winkleman JW (Eds). Chemical Chemistry: Principles and Techniques, 2nd edn. Harper and Row, Penncylvania; 1974.
2. Owen JA, Iggo B, Scandrett FJ, et al. The determination of creatinine in plasma or serum and in urine, a critical examination. Biochem J. 1954;58:426-37.

CHAPTER

15

Determination of Total Protein and Albumin

INTRODUCTION

Proteins are polymers of L-amino acids. There are numerous proteins in our body with different functions. Here we shall discuss plasma/serum proteins. Plasma and serum are both fluid parts of the blood. **Plasma** is the supernatant, obtained upon addition of an anticoagulant to the blood whereas **serum** is the supernatant obtained when the plain blood specimen is allowed to clot. Composition of serum is slightly different from plasma because it is lacking fibrinogen, clotting factors II, VII, VIII but has higher serotonin content due to the breakdown of platelets during clotting.

Plasma/serum proteins comprise a complex mixture of different proteins. The important proteins present in the plasma/serum are:
- Albumin
- Globulin
- Conjugated proteins such as lipoproteins
- Fibrinogen (absent from serum).

Biological Functions of Proteins

- Maintenance of oncotic pressure—albumin
- Transport of molecules—albumin transport bilirubin, transferrin transport iron
- Hormone function—TSH
- Coagulation—fibrinogen
- Defense—antibodies
- Nutritional—albumin
- Catalytic—enzyme proteins.

Liver is involved in the synthesis of albumin, fibrinogen, prothrombin, other clotting factors and several other proteins coming under alpha and beta globulins.

METHODS OF PROTEIN ESTIMATION

- Biuret method
- Direct photometric method
- Dye binding methods [using dyes like Amido black 10 B → electrophoretic band based quantitation and serum albumin estimation; Coomassie brilliant blue (CBB) → CSF protein estimation; Bromocresol green (BCG) → serum albumin; Bromocresol purple (BCP) → serum albumin]
- Lowry (Folin-Ciocalteu) method (for fibrinogen and total proteins)
- Kjeldahl's method (time consuming but still used to define reference proteins for biuret method)
- Refractometry (for getting a rapid estimate of total proteins)
- Turbidimetric and nephelometric methods.

Biuret Method[1]

Specimen

Serum or plasma.

Principle

Proteins form purple colored complex with cupric ions in alkaline solution.

The biuret test is given by those substances containing two carbamyl groups (CONH) joined either directly or by a single nitrogen or carbon atom. The purplish violet color is due to the formation of a copper coordination complex (see Biuret Test in Chapter – Reactions of Proteins). The molecule should have a minimum of two peptide bonds to give copper coordination complex that impart **violet** color to test mixture.

Reagents Required

- Sodium chloride 0.9% (Normal saline)
- Biuret reagent (see Preparation Reagent)
- Working Protein standard: (0.05 g/mL) (see Preparation Reagent).

Procedure

Set tubes and labeled as **T, S and B**. For further steps see Figure 15.1.

Calculation

Volume of (0.05 g/mL) standard solution taken = 100 µL
Amount of protein in 100 µL of standard (0.05 g/mL) solution = 0.005 g%
Concentration of total protein in 100 mL serum (g %) = T/S × concentration of standard × 100/volume of serum
= T/S × 0.005 × 100/0.1
= **T/S × 5 g%**

DETERMINATION OF ALBUMIN[2]

BCG Method (Bromocresol Green)[2]

Specimen: Serum is preferred.

Principle: Albumin binds with BCG dye at pH 4.15 giving rise to green color which shows maximum absorbance at 630 nm. Other proteins will not bind with this dye and hence do not interfere with this reaction.

Reagents

- BCG working agent
- Standard albumin solution (5 g/dL)
- Distilled water.

Procedure

- Take three test tubes and mark **T, S and B** on them.

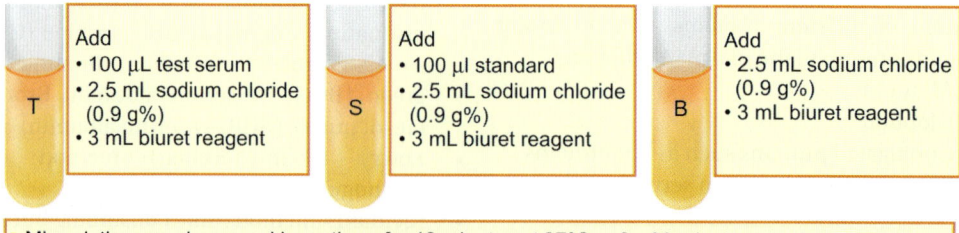

Figure 15.1 Procedure for total protein estimation by biuret method

- Add **5 mL BCG reagent** to all tubes.
- Add **20 μL (0.02 mL) distilled water** to B, **20 μL serum** to T and **20 μL standard** to S tubes. Mix well. Keep for 30 seconds and read using red filter or at 628 nm.
- Set the reading to zero absorbance with water. Then read absorbance with BCG reagent and reset the reading to zero against the BCG reagent. Then take the absorbance readings of Test and Standard.

Calculation

Volume of (0.05 g/mL) standard solution taken
$$= 100\ \mu L$$
Amount of Albumin in 0.02 mL of standard (0.05 g/mL) solution = 0.001 g%
Concentration of albumin in 100 mL serum (g%)
= OD of T/OD of S × Concentration of standard × 100/volume of serum taken
= T/S × 0.001 × 100/0.02 g%
= **T/S × 5 g%**

Reference Range

Serum total protein: 6.3–7.9 g%

Serum albumin: 3.7–5.3 g%

Serum globulins can be calculated from total protein and albumin values.

Serum globulins = Total protein–Albumin
Albumin Globulin ratio: 1.2 – 2.5: 1.

Interpretation

Total proteins include albumin, globulin and fibrinogen.

Serum Total Proteins
- Increased in
  - Dehydration
- Decreased in
  - Over hydration.
  - Cases with low albumin accompanied by no increase in globulin.

Serum Albumin
Decreased in:
- Loss of albumin, e.g. Nephritic syndrome, protein loosing enteropathy, burns, severe hemorrhage.
- Malabsorption of protein from the alimentary tract, e.g. malignancies of stomach, intestines and pancreas, enteritis.
- Decreased synthesis in liver diseases, e.g. cirrhosis.
- Increased catabolism of proteins (negative nitrogen balance), e.g. shock due to any cause, febrile illness, untreated diabetes mellitus, hyperthyroidism.

Serum Globulins
Increased in:
- Advanced liver disease
- Multiple myeloma
- Chronic infections, e.g. rheumatoid arthritis and tuberculosis
- Macroglobulinemia.

Albumin Globulin Ratio
Albumin globulin ratio—1.2 – 2.5: 1.

Note: A decrease in albumin and a rise in globulin may give low A:G ratio, in which case total protein remains within normal limits.

A:G ratio reversal seen in cases where albumin is low, e.g. chronic liver diseases like cirrhosis or cases in which globulins are produced excessively.

QUESTIONS

1. Name the method used in the estimation of total protein.
2. Name the serum protein fractions.
3. Name a protein fraction absent from serum but present in plasma.
4. Give the reference ranges of serum total protein and albumin.
5. How will you approximately find out the globulin concentration from serum total protein and albumin values?

6. Name the conditions in which total protein levels are low.
7. Name four conditions in which albumin levels are low.
8. What is A: G ratio and reversal of A:G ratio? Explain the significance.

REAGENT PREPARATION

- **Sodium chloride 0.9% (Normal saline):** Dissolve 9 g NaCl in a few mL of distilled water in a measuring cylinder or volumetric flask and make up to 1000 mL. This is stable at room temperature for 5–6 months.
- **Biuret reagent:** Dissolve **4 g NaOH** in 400 mL of distilled water. Add **4.5 g of sodium potassium tartrate** and mix to dissolve. Then add **1.5 g cupric sulfate pentahydrate ($CuSO_4\ 5\ H_2O$)** followed by **4.5 g potassium iodide**. Transfer the solution into a 500 mL volumetric flask or measuring cylinder and make up to 500 mL with distilled water. Keep in a brown bottle at room temperature. It is stable up to 6 months.
- **Protein standard stock (10 g%):**
    - Human serum pools are not recommended due to the risks of hepatitis B and HIV.
    - Lyophilized (freeze dried) protein standards are available commercially. But it is costly.
    - It can be prepared from less costly dried bovine albumin.
    - Weigh about 5.3 g (a little excess of wanted quantity) of bovine albumin powder. Dry it overnight in an oven at 60°C. Then from this dried powder, weigh out 5 g and add this into a beaker containing 25–30 mL of normal saline (NaCl 0.9 g%). Stir gently to dissolve it. Then transfer it to a **standard flask of 50 mL capacity**. Then makeup the volume to 50 mL with saline. This gives a protein standard of **10 g% strength**. This standard solution is stable for 6 months at 2–8°C.
- **Working protein standard (0.05 g/mL):** Pipette out 5 mL of stock standard into a 10 mL standard flask and make up to 10 mL with normal saline. This can be used as working standard for total protein and albumin estimations.
- **BCG reagent:** Dissolve 105 mg of BCG (3.3′, 5, 5′ tetrabromo-m-cresolsulfonphthalein) in 950 mL of water. Adjust the pH of the solution to 4.15–4.25 with NaOH (6 mol/L) and make up to 1 L with distilled water. Store in polyethylene bottle and store at room temperature. Stable up to 6 months.

REFERENCES

1. Kingsley GR. The direct biuret method for the determination of serum proteins as applied to photoelectric and visual colorimetry. J Lab Clin Med. 1942;27:840-5.
2. Silverman LM, Christenson RH, Griger H, et al. Amino acids and proteins. In: Tietz NW (Ed). Textbook of Chemical Chemistry. 1st edition. WB Saunder's Company; 1986.pp.588-9.

CHAPTER 16

Determination of Total Cholesterol

CHOLESTEROL

Cholesterol is a sterol with an alcoholic group. It is a tetracyclical compound containing cyclopentanoperhydrophenanthrene ring (Fig. 16.1). It is found in all types of cells.

Cholesterol in the body is derived from exogenous (diet) and endogenous source. Several physiologically important compounds are derived from it, e.g. vitamin D, bile acids, steroid hormones.

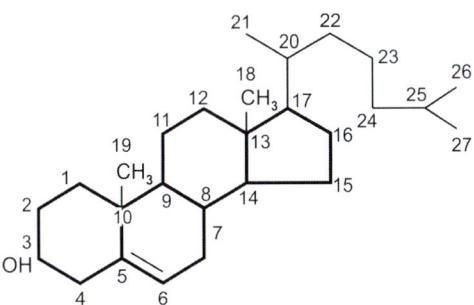

Figure 16.1 Structure of cholesterol

Methods of Total Cholesterol Estimation

- **Chemical methods**
  - Zak's method using reaction with ferric chloride and sulfuric acid
- **Enzymatic methods**
  - Cholesterol esterase, cholesterol oxidase and peroxidase based method (widely used).

Zak's Method[1] of Cholesterol Estimation using Reaction with Ferric Chloride and Sulfuric Acid (Fig. 16.2)

Specimen
Serum separated from plain blood collected in a dry bottle or blood collection tube.

Principle
Serum is treated with ferric chloride-acetic acid reagent to precipitate the proteins. The protein free filtrate is treated with sulfuric acid and acetic acid. The cholesterol present in the

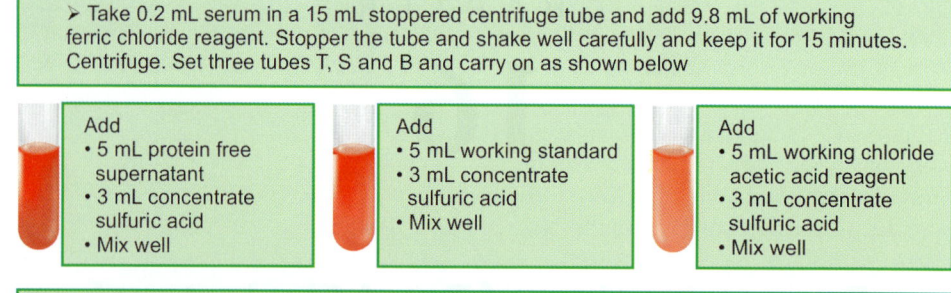

Figure 16.2 Procedure of Zak's method of cholesterol estimation

protein free filtrate is oxidized and dehydrated by ferric chloride, acetic acid and sulfuric acid to a **red** colored compound. The intensity of the color is proportional to the cholesterol content in the serum. It is read at 540 nm (green filter).

Reagents Required
- Ferric chloride acetic acid reagent
- Concentrated sulfuric acid (analytical grade)
- Cholesterol working standard (0.04 mg/mL).

Calculation
Concentration of standard in 5 mL standard solution = 5 × 0.04 mg/mL = 0.2 mg

0.2 mL serum diluted to 10 mL by adding 9.8 mL of ferric chloride acetic acid reagent, so that 1 mL contains 0.02 mL serum. Out of this, 5 mL used for the reaction.
∴ Volume of serum taken = 0.02 × 5 = 0.1
Serum cholesterol in 100 mL serum (mg %)
= Reading of test/Reading of standard × concentration of standard × 100/volume of serum taken
= T/S × 0.2 × 100/0.1 = T/S × 200 mg %

Enzymatic Method using Cholesterol Ester Hydrolase, Cholesterol Oxidase and Peroxidase

Specimen
Serum separated from **fasting** plain blood collected in a dry bottle or blood collection tube

Principle
Cholesterol esterase is used to free the cholesterol from cholesterol esters. The free cholesterol is oxidized by cholesterol oxidase producing hydrogen peroxide (H_2O_2) which gives a red color on reacting with phenol and 4 aminoantipyrine.

Cholesterol ester hydrolase
Cholesteryl ester + H_2O → Cholesterol + Free fatty acid.

Cholesterol oxidase
Cholesterol + O_2 → Cholestenone + H_2O_2

Peroxidase
H_2O_2 + Phenol + Chromogen → Red dye + H_2O

Procedure

About 3–10 mL of serum or plasma added to single reagent containing all enzymes and other ingredients and incubated under controlled conditions as specified in the brochure provided with the commercial reagent kits. The pink color developed is read **at 510 nm**.

Interpretation

Desirable level suggested by National Cholesterol Education Programme (NCEP) of total cholesterol
  Adults < 200 mg%
  Children and adolescents < 170 mg%
  Alterations in total cholesterol level are given in Table 16.1.

QUESTIONS

1. Give the importance of cholesterol in the body.
2. Describe the structure of cholesterol.
3. Name three biologically important compounds derived from cholesterol.
4. Name two food items rich in cholesterol.
5. Mention two methods by which serum total cholesterol can be estimated.
6. Give the principle of Zak's method estimating serum cholesterol.
7. Give the principle of enzymatic method of total cholesterol estimation.
8. Give the desirable serum cholesterol level recommended by NCEP in adults, adolescents and children.
9. What are the major causes of hypercholesterolemia?
10. What are the major causes of hypocholesterolemia?

TABLE 16.1 Alterations in total cholesterol

Hypercholesterolemia	Hypocholesterolemia (uncommon)
Diabetes mellitus	Hyperthyroidism
Nephrotic syndrome	Anemias
Obstructive jaundice	Hemolytic jaundice
Hypothyroidism	Malabsorption syndrome
Hypopituitarism— small increase	Severe wasting
	Acute infections and terminal states

REAGENT PREPARATION

- **Ferric chloride acetic acid reagent:** Dissolve 0.05 g ferric chloride ($FeCl_3 \cdot 6H_2O$) in 100 mL of analytical grade glacial acetic acid in a graduated cylinder.
- **Concentrated sulfuric acid (analytical grade).**
- **Cholesterol standard stock:** Dissolve 100 mg cholesterol in 100 mL glacial acetic acid in a standard flask (100 mL).
- **Working cholesterol standard:** Dilute the stock standard 1 to 25 with ferric chloride acetic acid reagent (0.04 mg/mL).

REFERENCE

1. Zak B. Cholesterol methodologies: A review. Clin Chem. 1977;23:1201.

CHAPTER 17

Determination of Uric Acid

URIC ACID

Uric acid is the major end product of catabolism of purine bases—adenine and guanine nucleotides of cellular DNA and RNA (endogenous). It is also formed from dietary nucleic acids (exogenous) (Fig. 17.1). Uric acid from endogenous source constitutes about 400 mg and from exogenous source it is about 300 mg.

Uric acid in the blood is filtered at the glomerulus and fully reabsorbed in the proximal tubule. The uric acid secreted in the distal convoluted tubule is partly reabsorbed partly excreted in urine.

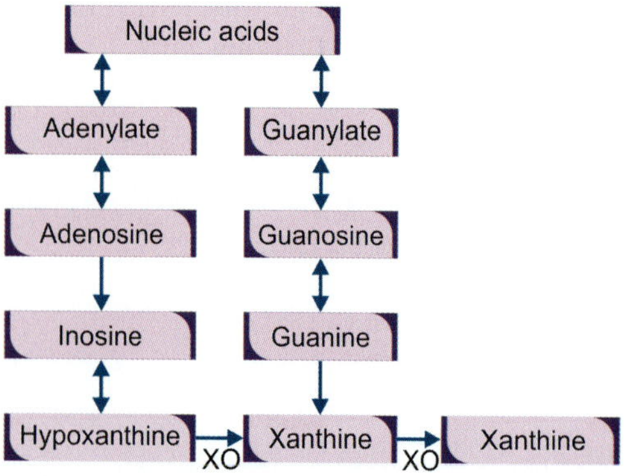

An understanding of solubility characteristics of uric acid is important to know the uric acid crystallization and stone formation. The first pka (dissociation constant) of uric acid is 5.57. The second pka is at 9.8 which do not come in the pH range of any physiological significance. *Above the pH 5.57, uric acid exist in ionized form, 'urate' ion which is more soluble than the unionized form (uric acid).*

Below the urine pH 5.75, it exists mainly in unionized form, 'uric acid' which is insoluble and tend to crystallize, when the concentration of it crosses saturation points in body fluids.

Methods of Uric Acid Estimation

- **Chemical methods**
  - Phosphotungstic acid method
- **Enzymatic method**
  - Uricase
- **HPLC** (High pressure liquid chromatography) method.

Figure 17.1 Formation of uric acid

Phosphotungstic Acid Method

Specimen
Serum, plasma (collected using the anticoagulant oxalate).

Principle
Uric acid is oxidized to allantoin and carbon dioxide by a phosphotungstic acid reagent in alkaline medium and phosphotungstic acid is in turn reduced to **tungsten blue** in the reaction. The intensity of the color developed is measured at wavelengths of **650–700 nm** in a spectrophotometer or by using red filter in a photoelectric colorimeter. Protein free filtrate is to be used to avoid turbidity and the quenching of the absorbance.

Reagents Required (For details see Preparation Reagent Preparation)
- Sodium tungstate 10 g/dL
- Sulfuric acid (0.33 mol/L)
- Phosphotungstic acid reagent
- Sodium carbonate (14 g/100 mL)
- Uric acid working standard (0.005 mg/mL).

Procedure
Preparation of protein free filtrate: Mix 1 mL of serum or plasma with 8.0 mL of distilled water, 0.5 mL of 0.33 molar H_2SO_4 and 0.5 mL of sodium tungstate (10 g%) in a tube and filter. (1 mL of this mixture contains 0.1 mL of serum).

Set tubes T, S and B for test, standard and blank respectively.
Procedure as shown in Figure 17.2.

Calculation
Concentration of uric acid in 100 mL blood (mg%)
= OD of T/OD of S × concentration of standard × 100/vol of serum taken
= T/S × 0.015 × 100/0.3
= T/S × 5 mg%

Interpretation
Reference range serum uric acid
Males: 4.4–7.6 mg% (0.26–0.45 mmol/L)
Females: 2.3–6.6 mg% (0.13–0.39 mmol/L)

(To convert mg% value to m mol/L, multiply mg % value with 0.059, e.g. 4.4 mg% × 0.059 → 0.26 mmol/L).

The level of uric acid gradually increases with age in both sexes especially after menopause in women. Those with serum uric acid levels more than 9.0 mg% are more prone for developing gouty arthritis.

Rate of uric acid excretion in individuals with unrestricted purine diet is 250–750 mg per day.

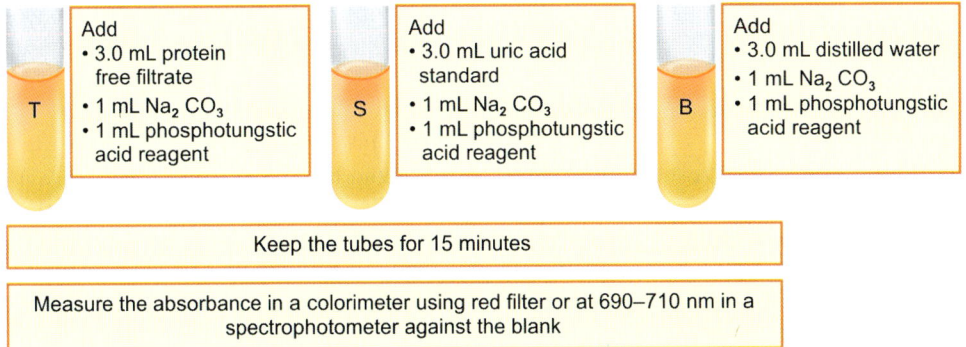

Figure 17.2 Procedure of phosphotungstic acid method of uric acid estimation

This may decrease to 400 mg/day upon a purine free diet. That is the importance of restriction of purine rich foods in cases of hyperuricemia.

Hyperuricemia

It is defined by serum or plasma uric acid levels greater than 7 mg% in men or greater than 6.0 mg% in women.

Causes of Hyperuricemia

Increased Formation
- Primary causes
  - Inherited metabolic disorders, e.g. Lesh Nyhan syndrome
- Secondary causes
  - Excess dietary intake
  - Increased nucleic acid turnover, e.g. malignancy, psoriasis.

Decreased Excretion
- Primary (idiopathic)
- Secondary causes
  - Chronic renal failure
  - Lactic acidosis
  - Thiazide diuretics therapy.

Hypouricemia

It is defined by serum or plasma uric acid levels less than 2 mg%.

It is rare where as hyperuricemia is common.

Causes

- Severe hepatocellular disease with reduced synthesis of purines.
- Defective renal tubular reabsorption of uric acid, e.g. Fanconi's syndrome congenital or acquired.
  - Acquired renal tubular damage due to toxic agents like radiopaque contrast media, cancer chemotherapy, over treatment with allopurinol.

QUESTIONS

1. Name the method used in the estimation uric acid in the serum. Give its principle.
2. Give the normal values of serum uric acid in males and females.
3. What are the factors affecting serum uric acid level in a normal person?
4. What is hyperuricemia? What are the different causes of it?
5. Define hypouricemia. Name the condition in which it is seen.
6. What is the rationale of giving alkalizer in patients with uric acid calculi.
7. What is gout? What do you mean by tophi?
8. Give the reason for getting high serum uric acid levels in patients with malignancy?
9. Name some purine-rich foods, the intake of which has to be restricted in patients with hyperuricemia.
10. Name one drug used to treat hyperuricemia that act at the level of xanthine oxidase. Describe its mechanism of action.

REAGENT PREPARATION

- **Phosphotungstic acid reagent:** Weigh 40 g of molybdenum free sodium tungstate AR and dissolve in 250–00 mL of distilled water. Slowly add concentrated 32 mL of pure ortho phosphoric acid cautiously.

  Reflux gently for 4 hours. Cool to room temperature. Add 300 mL distilled water. Add 32 g of lithium sulfate monohydrate into this. Mix and make up to 1 L. Store in a refrigerator.

- **Sodium tungstate (10 g/dL):** Take 10 g of sodium tungstate ($Na_2WO_4 2H_2O$) AR in a volumetric flask and dissolve in a few mL of distilled water and make up to 100 mL.

- **Sodium carbonate (14 g/100 mL):** Weigh 70 g of anhydrous sodium carbonate AR and add it to a few mL of water in a beaker and dilute to 500 mL. Transfer it to a polyethylene bottle.

- **Uric acid stock standard (100 mg/dL):** Weigh accurately 100 mg of uric acid AR and 60 mg of lithium carbonate AR (Li_2CO_3) and add them into a beaker. Add a few mL of distilled water and warm gently to dissolve the solids added. Cool and transfer with washing to a volumetric flask and make up to 100 mL with distilled water. It is stable for many months if refrigerated.
- **Working uric acid standard (0.005 mg/mL):** Pipette 0.5 mL of uric acid stock standard and add it into a 100 mL standard flask and Make up to 100 mL using distilled water. It is stable for 2–3 days if refrigerated.

CHAPTER

18

Determination of Bilirubin

INTRODUCTION

Bilirubin is an orange yellow pigment derived from heme. Daily bilirubin production is approximately 250–300 mg in humans from all sources. (About 85% of heme released from senescent RBCs in the reticuloendothelial system, and 15% from RBC precursors destroyed in the bone marrow and from catabolism of other heme containing proteins such as peroxidases, cytochromes and myoglobin).

Bilirubin is bound to albumin and transported to the liver. Inside the hepatocytes bilirubin is conjugated with glucuronic acid by UDP glucuronyl transferase to produce bilirubin glucuronides which then are excreted in bile into the intestine. In the intestine, bilirubin glucuronides are hydrolyzed by bacterial β glucuronidase to form deconjugated bilirubin which is then reduced by anaerobic intestinal microorganisms to form colorless urobilinogens which includes urobilinogen, stercobilinogen, mesobilinogen. About 20% of urobilinogens produced are reabsorbed from intestine and enters enterohepatic circulation. About 2–5% of this enters the systemic circulation and appears in urine as urobilinogen (Fig. 18.1). Stercobilinogen along with unabsorbed urobilinogen and mesobilinogen are oxidized in the lower intestine to form orange brown colored stercobilin, mesobilin and urobilin and excreted in feces.

Biliverdin is formed by mild oxidation of bilirubin. It is formed spontaneously when bilirubin is oxidized by exposure to air in alkaline solution or oxidized with ferric chloride in acetic acid or is treated with H_2O_2. It is dark green in color.

Urobilins are formed by mild oxidation of urobilinogen on exposure to air or to mild oxidizing agents. Urobilins are reddish orange in color. It is formed readily when urine is exposed to light and air and its formation is slow at alkaline pH.

METHODS

- Using diazo reagent
  - Method of Malloy and Evelyn
  - Method of Powell
  - Method of Jendrassik and Grof
- Direct spectrophotometry ($A_{454} - A_{540}$) in infants.

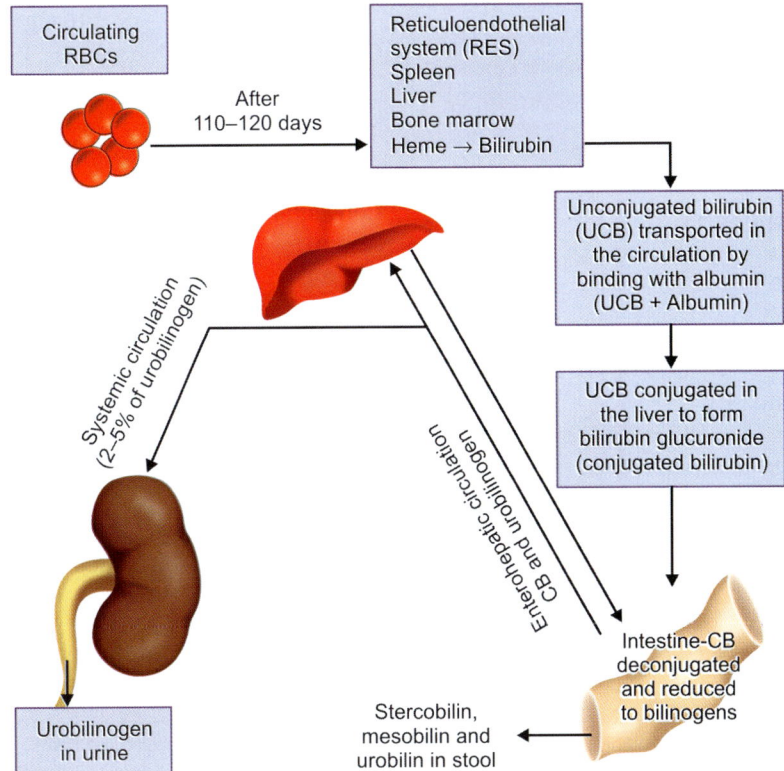

Figure 18.1 Formation and fate of bilirubin

Method of Malloy and Evelyn for the Determination of Serum Bilirubin[1]

Specimen

Fresh Serum or Plasma.

Principle

It is based on the formation of purple azobilirubin when bilirubin reacts with diazo reagent introduced by Van den Berg hence called **Van den Berg reaction**. Bilirubin in circulation is of two types—unconjugated bilirubin (UCB) bound to albumin and conjugated bilirubin (CB) not bound to albumin.

Van den Berg reaction consists of two types of reactions:

Direct Van den Berg reaction: The conjugated bilirubin not bound to albumin (water-soluble bilirubin glucuronides) reacts immediately with diazo reagent to give purple azobilirubin.

Indirect Van den Berg reaction: The unconjugated bilirubin bound to albumin, reacts very slowly and requires an accelerator like methanol. Methanol releases albumin bound to bilirubin and exposes the carboxyl groups of it to diazo reagent.

Total bilirubin value (conjugated + unconjugated bilirubin) given by direct and indirect reaction. The intensity of the color developed is directly proportional to the concentration of the bilirubin in the serum and is read at **540 nm**.

Reagents (for details see Reagent Preparation)
- Diazo reagent
- 1.5% HCl
- Bilirubin standard: 0.1 mg/mL
- Absolute methanol.

Procedure (Fig. 18.2)

Calculation

OD of Dt (Direct bilirubin test (T_{DB})
 = OD of Dt–OD of Db
OD of standard (S_B)
 = OD of S–OD of B
Concentration of standard in 0.2 mL
 = 0.02 mg
Concentration of Dt (Direct bilirubin in test serum)
 = T_{DB}/S_B × concentration of standard × 100/volume of serum
 = T_{DB}/S_B × 0.02 × 100/0.2 mg%
 = T_{DB}/S_B × 10 mg%

OD of Tt (Total bilirubin test) (T_{TB})
 = OD of Tt–OD of Tb
OD of standard (S_B)
 = OD of S–OD of B (same as above)
Concentration of Tt (Total bilirubin in test serum)
 = T_{TB}/S_B × concn of std × 100/volume of serum
 = T_{TB}/S_B × 0.02 × 100/0.2 mg%
 = T_{TB}/S_B × 10 mg%

INTERPRETATION

Reference Range

Total bilirubin (TB): 0.2–1.0 mg%.
Direct bilirubin (conjugated bilirubin) (CB): 0.2–0.5 mg%.
Unconjugated bilirubin (UCB) = Total bilirubin– Conjugated (direct) bilirubin.

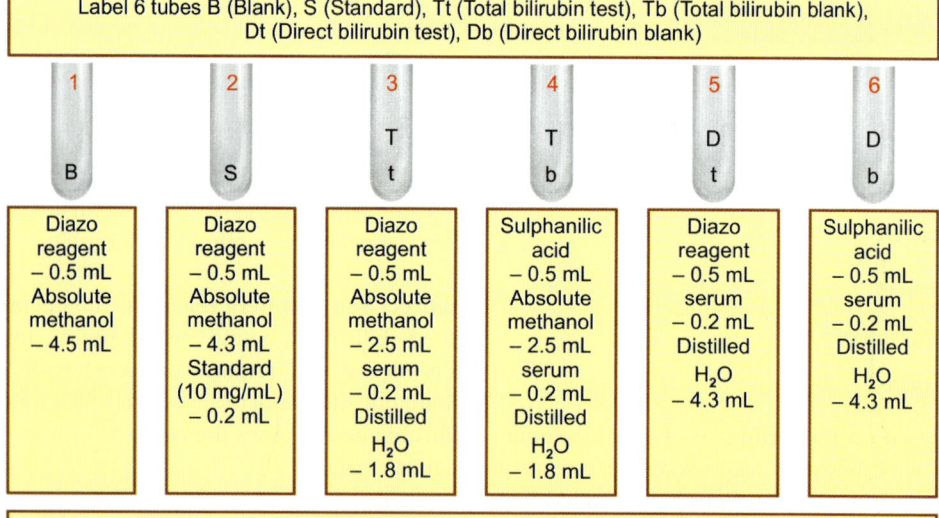

Figure 18.2 Procedure of bilirubin estimation (Modified method of Malloy and Evelyn)

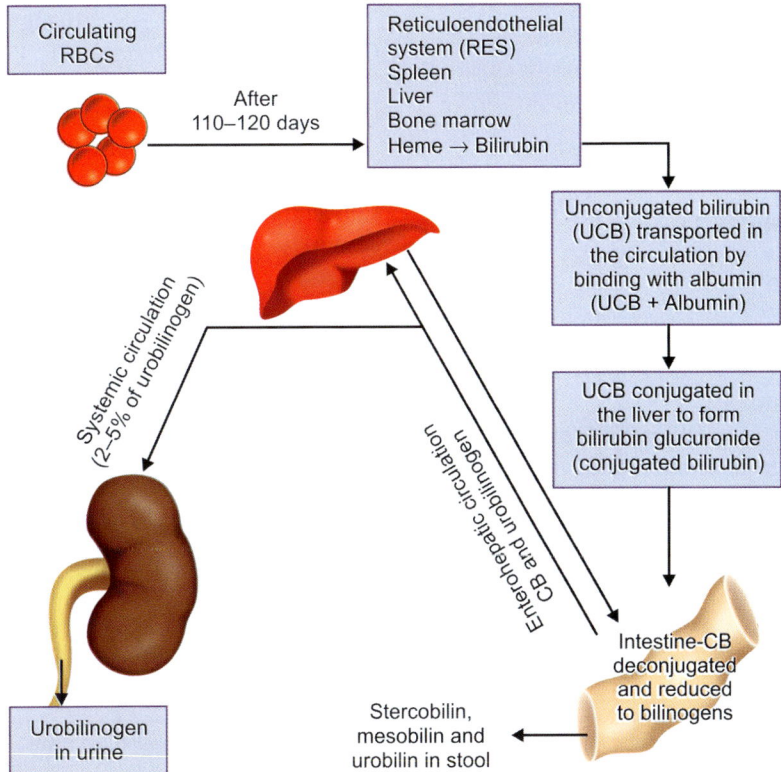

Figure 18.1 Formation and fate of bilirubin

Method of Malloy and Evelyn for the Determination of Serum Bilirubin[1]

Specimen

Fresh Serum or Plasma.

Principle

It is based on the formation of purple azobilirubin when bilirubin reacts with diazo reagent introduced by Van den Berg hence called **Van den Berg reaction**. Bilirubin in circulation is of two types—unconjugated bilirubin (UCB) bound to albumin and conjugated bilirubin (CB) not bound to albumin.

Van den Berg reaction consists of two types of reactions:

Direct Van den Berg reaction: The conjugated bilirubin not bound to albumin (water-soluble bilirubin glucuronides) reacts immediately with diazo reagent to give purple azobilirubin.

Indirect Van den Berg reaction: The unconjugated bilirubin bound to albumin, reacts very slowly and requires an accelerator like methanol. Methanol releases albumin bound to bilirubin and exposes the carboxyl groups of it to diazo reagent.

Total bilirubin value (conjugated + unconjugated bilirubin) given by direct and indirect reaction. The intensity of the color developed is directly proportional to the concentration of the bilirubin in the serum and is read at **540 nm**.

Reagents (for details see Reagent Preparation)
- Diazo reagent
- 1.5% HCl
- Bilirubin standard: 0.1 mg/mL
- Absolute methanol.

Procedure (Fig. 18.2)

Calculation

OD of Dt (Direct bilirubin test (T_{DB}))
= OD of Dt−OD of Db
OD of standard (S_B)
= OD of S−OD of B
Concentration of standard in 0.2 mL
= 0.02 mg
Concentration of Dt (Direct bilirubin in test serum)
= T_{DB}/S_B × concentration of standard × 100/volume of serum
= T_{DB}/S_B × 0.02 × 100/0.2 mg%
= T_{DB}/S_B × 10 mg%
OD of Tt (Total bilirubin test) (T_{TB})
= OD of Tt−OD of Tb
OD of standard (S_B)
= OD of S−OD of B (same as above)
Concentration of Tt (Total bilirubin in test serum)
= T_{TB}/S_B × concn of std × 100/volume of serum
= T_{TB}/S_B × 0.02 × 100/0.2 mg%
= T_{TB}/S_B × 10 mg%

INTERPRETATION

Reference Range

Total bilirubin (TB): 0.2−1.0 mg%.
Direct bilirubin (conjugated bilirubin) (CB): 0.2−0.5 mg%.
Unconjugated bilirubin (UCB) = Total bilirubin−Conjugated (direct) bilirubin.

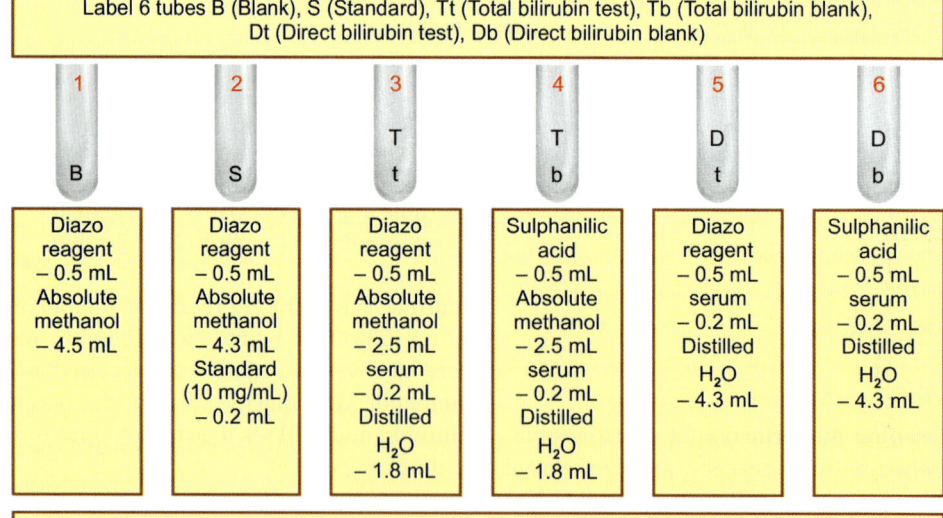

Figure 18.2 Procedure of bilirubin estimation (Modified method of Malloy and Evelyn)

JAUNDICE

Defective bilirubin metabolism or biliary obstruction results in raised bilirubin in blood and yellowish staining of skin, sclera and mucous membranes called jaundice (bilirubin >2 mg%).

Types of Jaundice

- **Prehepatic jaundice:** Excessive production, e.g. hemolytic jaundice—here erythrocytes undergo hemolysis excessively to produce heme and bilirubin in excess.
  Van den Bergh reaction: Indirect reaction
  Serum TB: ↑; Serum CB: Normal or slightly ↑; Serum UCB: Highly ↑
- **Hepatic jaundice:** Disorders of liver causing defective conjugation or defective secretion into bile, e.g. different kinds of hepatitis, glucuronyl transferase deficiency.
  Van den Bergh reaction: Biphasic (Both direct and indirect reaction due to increase in both CB and UCB).
  Serum TB: ↑; Serum CB: Moderate ↑; Serum UCB: Moderate ↑
- **Posthepatic jaundice:** Defective secretion due to obstruction of biliary pathways, e.g. biliary atresia, stones in common bile duct, Carcinoma head of pancreas pressing the common bile duct.
  Van den Bergh reaction: Direct reaction due to the predominance of CB.
  Serum TB: ↑; Serum CB: Highly ↑; Serum UCB: Normal or slightly ↑

QUESTIONS

1. Explain how bilirubin is formed in the body?
2. Describe the excretion of bilirubin from the body.
3. What is jaundice? Give three major types of jaundice.
4. How will you diagnose jaundice in terms of serum bilirubin?
5. How will you estimate bilirubin? What is the basis of this?
6. What is Van den Bergh reaction and comment on the types of van den Berg reaction?

REAGENT PREPARATION

- **Solution A:** Dissolve 1 g of sulfanilic acid in 15 mL of concentrated hydrochloric acid and make up to 1 liter with water. Keep in a brown bottle at room temperature. Stable for 6 months.
- **Solution reagent B (sodium nitrite):** Dissolve 0.5 g of sodium nitrite in water and make up to 100 mL with distilled water. Store at 2-8°C. Prepare freshly at frequent intervals (once in 3 weeks).
- **Diazo reagent:** Prepare freshly before use by mixing 0.3 mL solution B with 10 mL sulfanilic acid solution (solution A). After 2 minutes of mixing, it is ready to use. Keep at 2-8°C and use within 2-4 hours.
- **1.5% HCl (v/v):** 1.5 mL concentrated HCl made up to 100 mL with water.
- **Absolute methanol:** Dispense from the bottle.
- **Standard solution of bilirubin:** Dissolve 10 mg in 100 mL chloroform. Protect from light in a brown bottle. Keep at 4-6°C. The purity of bilirubin standards varies with suppliers. Hence, instead of true bilirubin methyl red standard can be used.

REFERENCE

1. Malloy HT, Evelyn KA. The determination of bilirubin with the photoelectric colorimeter. J Biol Chem. 1937;119:481-90.

CHAPTER

19

Determination of Transaminases

TRANSAMINASES

Transaminases catalyzes the transfer of amino group from an α-amino acid to an α-oxoacid leading to the formation of a different α-amino acid and a different α-oxoacid. All the primary alpha amino acids except (lysine, threonine, proline) can undergo such trasamination reactions catalyzed by different types of transaminases. Out of **these aspartate aminotransferase (AST) (EC 2.6.1.1)** (Former name Glutamate oxaloacetate transaminase—GOT) and **alanine aminotransferase (ALT) (EC 2.6.1.2)** (Former name Glutamate pyruvate transaminase—GPT) Reactions catalyzed by these enzymes are shown in the Figure 19.1.

Transaminases are present in most of the tissues in the body. They are the enzymes of cytoplasm. They are present in the plasma of healthy individuals in small amounts. During health, enzymes along with other molecules are retained in the cells by metabolically active plasma membrane. Integrity of the plasma membrane depends on the availability of cellular currency ATP. When the ATP synthesis is impaired due to deficiency of fuels (oxidizable substrates) or anoxia, cell membrane function (integrity) deteriorates. Molecules leak out of the cells. Cytoplasmic enzymes appear in plasma earlier than membrane bound enzymes.

Methods

- **Photometric methods**
  - Using 2, 4 Dinitrophenylhydrazine
  - By combining with diazonium salts.
- **Continuous monitoring methods**
  - Coupling transaminase reaction with specific dehydrogenate reaction.

Assay of Alanine Aminotransferase by Reitman and Frankel Method using 2, 4 Dinitrophenylhydrazine (Colorimetric Method)[1]

Principle

Alanine aminotransferase (ALT) catalyze transamination of L-alanine with α-ketoglutarate to form L-glutamate and pyruvate using pyridoxal phosphate as coenzyme.

To estimate ALT activity, serum is treated with alanine and alpha ketoglutarate (sub-

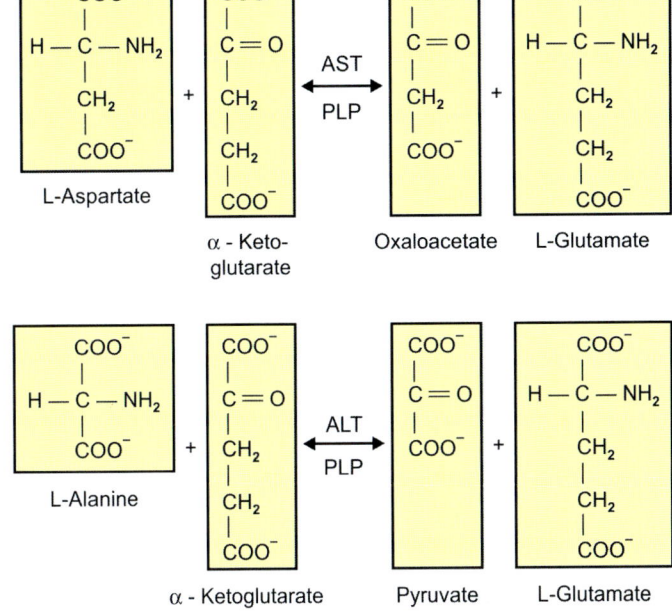

Figure 19.1 Reactions catalyzed by AST and ALT

strates). Then 2,4 dinitrophenylhydrazine added, which will couple with the pyruvate formed in the reaction in alkaline medium to form reddish brown hydrazone. The intensity of the color is proportional to the ALT enzyme activity in the serum and is measured at 505 nm in a spectrophotometer or with the green filter in a colorimeter.

Reagents (for details see Reagent Preparations)
- Phosphate buffer 0.1 M (pH 7.4)
- Buffered substrate for ALT
- Pyruvate standard (2 mmol/mL)
- Color reagent: 2, 4 DNPH (dinitrophenylhydrazine) 1 mmol/L
- NaOH 0.4 M.

Procedure (Fig.19.2)
Calculation: ALT activity in the serum μmol/min/L.

$$= \frac{\text{OD of T} - \text{OD of B}}{\text{OD of S} - \text{OD of B}} \times \text{Concentration of}$$

standard in μmol × 1000/vol of serum × 1/30 (incubation time)

$= T/S \times 0.4 \times 1000/0.2 \times 1/30 = T/S \times 66.7$ Cabaud units

Note: To convert Cabaud units to IU/L, multiply with factor 0.483.

Assay of Aspartate Aminotransferase by Reitman and Frankel Method using 2,4 Dinitrophenylhydrazine (Colorimetric Method)[1]

Same as that for ALT except for three differences:
1. For ALT, the substrates are DL Alanine and α-KG whereas for AST they are DL Aspartate and α-KG.
2. Incubation time is 30 minutes for ALT and it is 60 minutes for AST.

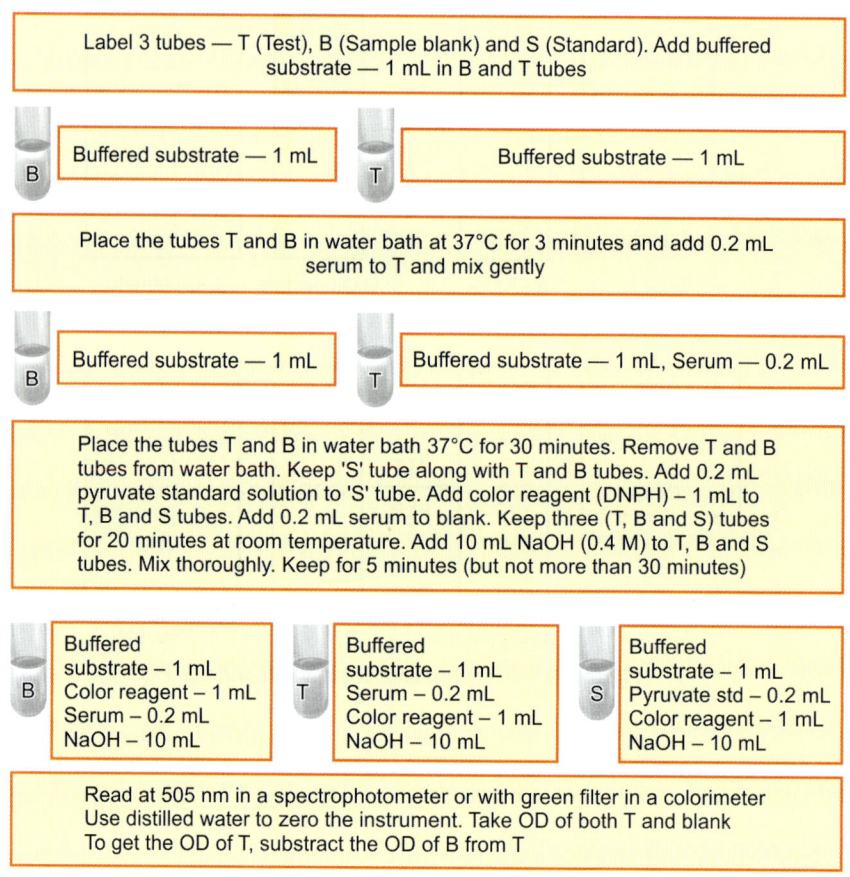

Figure 19.2 Procedure of ALT estimation

Principle

Aspartate aminotransferase (SGOT) catalyze transamination of L-aspartate with a-ketoglutarate to form L-glutamate and oxaloacetate (OA) using pyridoxal phosphate as coenzyme. The oxaloacetate formed undergoes spontaneous decarboxylation to form pyruvate. This pyruvate reacts with 2, 4 dinitrophenylhydrazine to give a yellow colored hydrazone which in alkaline medium forms reddish brown colour. To estimate AST activity, serum is treated with aspartate and alpha ketoglutarate (substrates).

Reagents (For details see Reagent Preparations)

- Phosphate buffer substrate – 0.1 M pH 7.4
- Buffered substrate for AST
- Pyruvate standard (2 mmol/mL)
- 2, 4 DHPH (dinitrophenylhydrazine) 1 mM/L
- NaOH 0.4 M.

Procedure (Fig.19.3)

Calculation: AST activity in the serum μmol/min/L

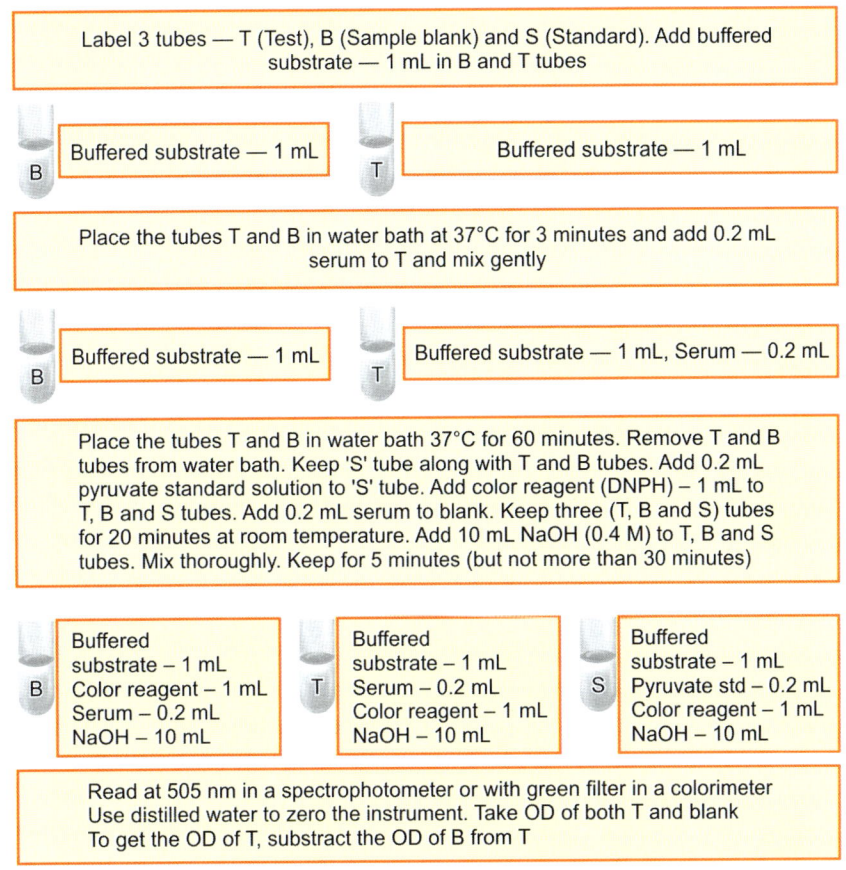

Figure 19.3 Procedure of AST estimation

$$= \frac{\text{OD of T} - \text{OD of B}}{\text{OD of S} - \text{OD of B}} \times \text{Concentration of}$$

standard in µmol × 1000/vol of serum × 1/60 (incubation time)

$= \text{T/S} \times 0.4 \times 1000/0.2 \times 1/60 = \text{T/S} \times 33.3$ Cabaud units

Note: To convert Cabaud units to IU/L, multiply with factor 0.483.

Points to Ponder
Collect blood in dry containers. Hemolyzed specimen will give falsely high AST and ALT activities.

TABLE 19.1 Reference interval of AST and ALT

Enzyme	Adult serum level
AST	8–20 U/L
ALT	10–40 U/L

Interpretation
Highest activity of AST (7800 times the normal serum level) is in the myocardium and next in the liver (7000 times the serum level) and next in skeletal muscle (5000 times). Whereas highest activity of ALT (2850 times the normal serum level) is in the liver and next in the kidney

(1200 times the serum level) and only 450 times the serum level in the myocardium and 300 times in skeletal muscle. In clinical practice both AST and ALT are assayed for diagnosing liver diseases and AST is used for evaluating ischemic heart disease occasionally. (See Table 19.1 for reference intervals of AST & ALT).

ALT and AST in Hepatic Disorders

- **Viral hepatitis and other types of liver diseases:** In hepatocellular diseases except viral hepatitis, transaminases are elevated to produce ALT/AST ratio less than 1. This ratio is known as **De Ritis ratio**. It becomes elevated to unity or greater than 1 in cases of infectious hepatitis and other types of inflammatory diseases of liver. In cirrhosis liver, the ratio is elevated slightly depending on the degree of hepatocellular necrosis. In terminal cirrhosis it is less than 1.
- **Primary and secondary carcinoma:** 5–10 times the normal activity of ALT and AST is observed.
- **Toxic hepatitis:** Very high (20 times) activity of both enzymes.

AST in Ischemic Heart Disease

Serum AST activity increase only 6–8 hours after the onset of chest pain and it peaks around 18–24 hours and fall to within the reference range by 4th–5th day in cases where no fresh infarct has been developed.

QUESTIONS

1. Describe the catalytical role of transaminases.
2. Name two tissues each in which high activity of AST and ALT are observed.
3. Give the reference ranges of serum AST and ALT in adults.
4. Give the principle of a method employed in the assay of serum transaminases.
5. What are the diagnostic uses of ALT and AST?
6. What is De-Ritis ratio? What is the application of using the De-Ritis ratio in evaluating different types of hepatic diseases?
7. Name one technique that will help in the separation of different isoenzyme fractions in the serum.
8. What is the pattern of rise of serum AST in myocardial infarction?
9. What is the precaution to be taken during the collection of blood for assay of transaminases?

REAGENT PREPARATION

- **Phosphate buffer – 0.1 M pH 7.4:** Dissolve **14.9 g disodium hydrogen phosphate dihydrate** (11.9 g of anhydrous disodium hydrogen phosphate) and **2.2 g of anhydrous potassium dihydrogen phosphate** in a few mL of distilled water and in a volumetric flask or cylinder or beaker and make up to 1000 mL with distilled water. Check the pH after adding 900 mL water and if the pH is less than 7.4 add a small amount of disodium hydrogen phosphate. If the pH is more than 7.4 add a pinch of potassium dihydrogen phosphate. It is stable for 2 months at 2–8°C.
- **Buffered substrate for ALT:** Dissolve **1.78 g DL alanine** and **30 mg alpha ketoglutaric acid** in **20 mL of phosphate buffer** (pH 7.4) and add 1–1.25 mL of 10% NaOH to adjust the pH to 7.4, in a beaker. Transfer the contents to a volumetric flask rinsing the beaker to the volumetric flask and make the volume to 100 mL with phosphate buffer. Adjust the pH to 7.4, by adding 10% NaOH in drops if needed. Add 1 mL chloroform as a preservative. And keep at 2–8°C. It is stable up to 2 weeks. Discard, if any turbidity develops.
- **Pyruvate standard (2 mmol/mL):** Dissolve 220 mg sodium pyruvate in phosphate buffer and make up to 100 mL with phosphate buffer in a volumetric flask. Transfer 10 mL of this into another volumetric flask and

dilute to 100 mL with phosphate buffer to get 2 mmol/mL pyruvate working standard. Discard the remaining 90 mL initial pyruvate solution. The working standard must be aliquoted into 5 mL sized containers in the freezer.
- **2,4 DHPH (dinitrophenylhydrazine) 1 mM/L:** Dissolve 200 mg DNPH in hot 1N HCl in a beaker. Allow to cool and make up to 1 L with 1 N HCl. Store at 2–8°C in amber colored bottles. Stable up to 6 months.
- **NaOH 0.4 M/L:** Dissolve 16 g NaOH in a little distilled water in a 1 L cylinder and make up to 1 L with distilled water. Keep in a stoppered polythene bottle at 25–30°C. Stable up to 6 months.
- **HCl 1N (1M/L):** Dilute 90 mL concentrated HCl to 1 L with distilled water in a graduated cylinder.

REFERENCE

1. Reitman S, Frankel S. A Colorimetric method for the determination of serum glutamic oxalacetic and glutamic pyruvic transaminases. Amer J Cli Path. 1957;28:56-63.

CHAPTER 20

Determination of Alkaline Phosphatase

INTRODUCTION

Alkaline phosphatase (ALP) is an enzyme that catalyze hydrolysis of monophosphoric esters to liberate phosphoric acid at alkaline pH (optimum pH 10). Several isoenzymes of ALP are recognized, e.g. those derived from liver, bones, intestine, kidney and placenta. Of this, the isoenzyme derived from the liver constitutes the major fraction in the serum normally. Next comes the one derived from skeleton, then intestine. Placental isoenzyme is produced during pregnancy.

Assay of serum alkaline phosphatase is useful routinely in the diagnosis of hepatobiliary disease and diseases of skeletal system associated with increased osteoblastic activity.

METHODS

- Method using 4-NPP (Nitrophenyl phosphate) (e.g. Modified method of Bowers and McComb).
- Method using disodium phenyl phosphate (e.g. King and Armstrong; Kind and King method).

Assay of Alkaline Phosphatase Kind and King Method (Fig. 20.1)[1]

Specimen

Serum or Heparinized plasma.

Principle

Disodium phenyl phosphate is acted upon by alkaline phosphatase in the serum and is hydrolyzed to release phenol and sodium phosphate. The phenol then reacts with 4-aminoantipyrine in the presence of potassium ferricyanide to give purple color and is read photometrically at 525 nm (green filter).

Reagents (For details see Reagent Preparations)

- Buffered substrate (Disodium phenyl phosphate in buffer)
- Phenol working standard 0.01 mg/mL
- Sodium hydroxide 0.5 N
- Sodium bicarbonate 0.5 M
- 4-aminoantipyrine 0.6%
- Potassium ferricyanide 2.4%

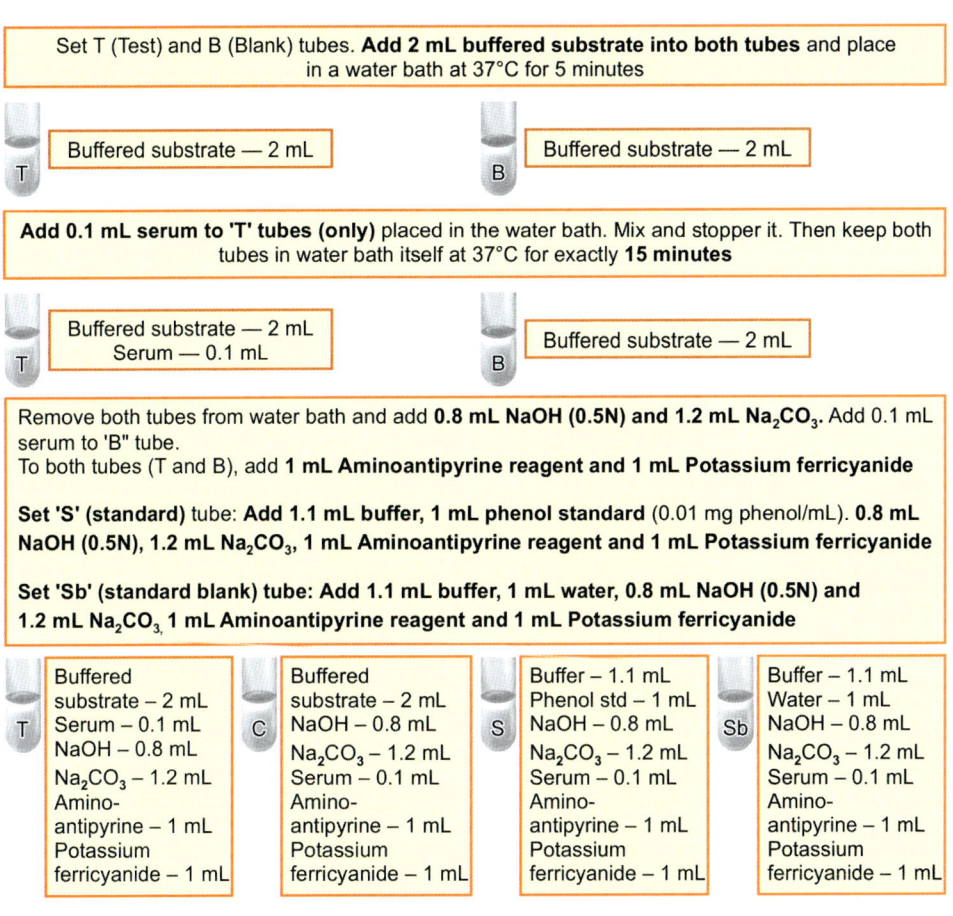

Figure 20.1 Procedure of Kind and King method of ALP estimation

- Sodium carbonate-sodium bicarbonate buffer (0.1 M).

Calculation

Serum alkaline phosphatase in King Armstrong units/100 mL (KAU%)

$$= \frac{\text{Reading of T} - \text{Reading B}}{\text{Reading of S} - \text{Reading of Sb}} \times$$

Concentration of standard × 100/volume of serum

= T/S × 0.01 × 100/0.1 = **T/S × 10**

To convert King Armstrong units/100 mL into IU/L: Multiply by 1000 and divide by 94 (MW of Phenol)

To convert 15 KAU % into IU/L = 15 × 1000/94 = 160 IU/L

Interpretation

Reference range: 40–125 IU/L.

High Alkaline Phosphatase Activity
- Physiologically
  - First few months of life
  - Pubertal growth spurt.

- Pathologically
  - Bone diseases: Where osteoblasts are active, i.e. when bone regeneration is taking place.
    - Rickets
    - Osteomalacia
    - Malabsorption of vitamin D and Calcium
    - Paget's disease (osteitis deformans)
    - Osteogenic sarcoma
    - Secondary deposits in bone
  - Diseases of liver and biliary tract
    - Markedly elevated in obstructive jaundice.

Points to Ponder: 5' nucleotidase activity level in the serum will help to find out the origin of high serum ALP levels in doubtful cases. 5' nucleotidase activity will be raised in ALP rise due to hepatic disease but will be normal in those with bone diseases.

ISOENZYMES OF ALKALINE PHOSPHATASE (ALP)

True isoenzymes are multiple forms of an enzyme catalyzing the same reaction but encoded by different genes producing different structure of enzymes. But many posttranslational modifications cause heterogeneity of various enzymes. Alkaline phosphatase (ALP) isoenzymes are due to both genetic and nongenetic modifications. Genetic loci of ALP are in chromosome 1 and 2. The locus of commonly encountered ALP isoenzymes (liver, bone, kidney and intestine) are located in chromosome 1 and that of placenta in chromosome 2. The former four have different degree of dilation (the number of sialic acid residues attached to the enzyme protein). These isoforms in general can be measured by different types of analytical techniques—electrophoresis, chromatography, chemical inactivation, immunochemical methods and methods based on differences in the catalytical properties of isoenzymes.

QUESTIONS

1. Describe the catalytic role of alkaline phosphatase.
2. What is the pH optimum of alkaline phosphatase?
3. Where is this enzyme located inside a cell?
4. What are the different isoenzymes of alkaline phosphatase?
5. Normally which isoenzyme predominate in the serum?
6. What is the role of alkaline phosphatase in the diagnosis of jaundice?
7. Name one technique that will help in the separation of different isoenzyme fractions in the serum.
8. Name one method by which you can determine the activity of ALP in the serum.
9. What is the formula for converting KAU units of ALP activity into IU/L?

REAGENT PREPARATION

- **Disodium phenyl phosphate 0.01 M:** Dissolve 1.09 g in distilled water and make up to 500 mL in a 750 mL – 1 L beaker. Heat to boil quickly, then cool, add a 1 mL chloroform and keep in a refrigerator.
- **Sodium carbonate–sodium bicarbonate buffer pH 10 (0.1 M):** Dissolve **3.18 g anhydrous sodium carbonate (Na_2CO_3)** and **1.68 g sodium bicarbonate ($NaHCO_3$)** in a few mL of water in a beaker and make up to 500 mL.
- **Buffered substrate for use:** Mix equal volumes of solutions 1 and 2. This will have a pH of 10.
- **NaOH 0.5 N:** Dissolve 20 g in 1 L of distilled water.
- **Sodium bicarbonate 0.5 N:** Dissolve 42 g in 1 L of distilled water.

- **4-aminoantipyrine 0.6%:** Dissolve 0.6 g in 100 mL of distilled water.
- **Potassium ferricyanide 2.4%:** Dissolve 2.4 g in 100 mL of distilled water.
- **Phenol stock standard (1 mg/mL):** Dissolve 1 g of pure crystalline phenol in 0.1 N HCl and make up to 1 L with acid. Store at 4°C.
- **Phenol working standard 0.01 mg/mL (1 mg/100 mL):** Dilute 1 mL stock phenol to 100 mL with water in a standard flask. Prepare whenever necessary, if any excess store at 4°C.

REFERENCE

1. Kind PRN, King EJ. Estimation of plasma phosphatase by determination of hydrolysed phenol with amino-antiphyrine. J clin Pathol. 1954;7(4):322-6.

CHAPTER 21

Determination of Total Calcium

CALCIUM

Calcium is found mainly in the skeleton and teeth. It is also present in plasma and other body fluids. In blood 50% of calcium is free, 40% protein bound and 10% is complexed with diffusible ions like bicarbonate, lactate, phosphate and citrate. About 32% of total blood calcium is albumin bound and 8% is bound to globulins. Calcium binds to negatively charged sites on proteins. The charge of proteins is pH dependent. For example, alkalosis leads to more basic pH which in turn causes increase in negative charge on proteins enhancing calcium binding. This lowers the level of free calcium in the blood. The reverse happens with acidosis. Thus, calcium is distributed among three plasma pools. Calcium in the plasma is redistributed among these three pools depending on protein concentrations, changes in pH and changes in free and total calcium concentrations in the serum.

Intracellular calcium participates in muscle contraction, hormone secretion, second messenger of hormone action, metabolic activities, enzyme actions, exocytosis and cell division.

A decrease in serum free calcium (either due to actual decrease or due to relative decrease caused by alkalosis) causes increased neuromuscular excitability and tetany whereas an increase in free calcium reduces neuromuscular excitability (Fig. 21.1).

Methods

- Photometric methods
  - OCPC (O-Cresolphthalein complexone) method
  - Arsenazo III method
- Ion selective electrodes (ISE)
- Atomic absorption spectrometry (reference method for serum total calcium)
- Titration method.

Determination of Calcium by Photometric Method: Ortho-Cresolphthalein Complexone Method [1]

Specimen

Serum/Plasma—use heparin as anticoagulant.

Points to Ponder
Factors leading to spurious rise in blood calcium:
- Use of tourniquet during blood collection
- Venostasis
- Erect posture.

Principle
In alkaline pH, O-cresolphthalein complexone (CPC) forms a complex (purple color with bluish tint) with calcium. The intensity of the color is a measure of concentration of calcium in the serum. The serum sample is diluted with acid to release protein bound and anion bound calcium. Interference with the divalent magnesium ions are reduced by addition of 8 hydroxyquinoline. It is read at 575 nm in a spectrophotometer or by using yellow filter in a colorimeter.

Reagents Required (For details see Reagent Preparations)
- Buffer
- Chromogen
- Color reagent
- Calcium standard 2.5 mmol/L
- EDTA.

Procedure (Fig. 21.2)

Calculation
OD of Test (T) = OD of test–OD of blank
OD of Standard (S) = OD of standard–OD of blank

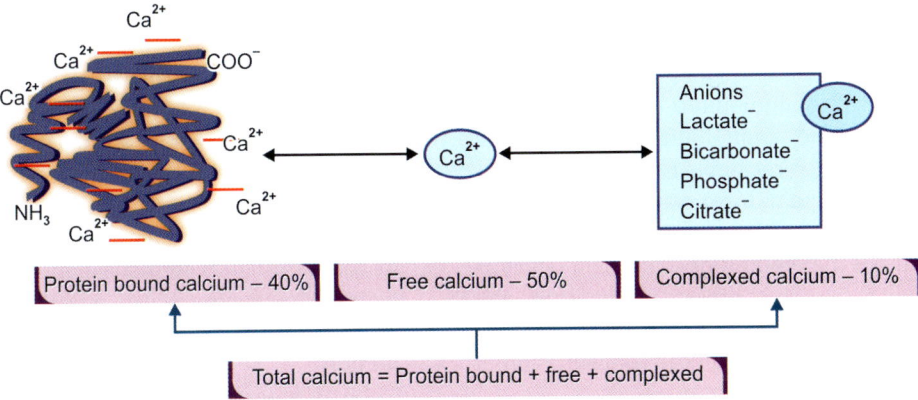

Figure 21.1 Three plasma pools of calcium

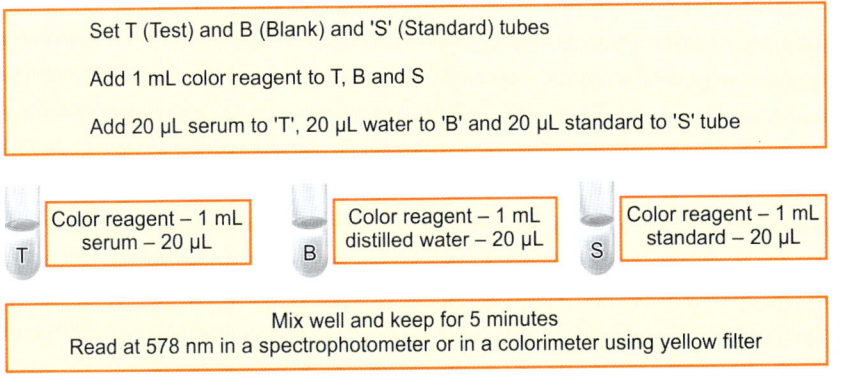

Figure 21.2 Procedure of CPC method of calcium estimation

Concentration of standard used
$$= 2.5 \text{ mmol/L}$$
Concentration of std in 20 µL
$$= (2.5/1000 \times 1000) \times 20$$
$$= 0.00005 \text{ mmol}$$
Volume of serum taken 20 µL
$$= 0.02 \text{ mL}$$
Concentration of calcium in mmol/L
$$= T/S \times 0.00005 \times 1000/0.02$$
$$= \mathbf{T/S \times 2.5}$$

Titration Method—Method of Clark and Collip (1925)[2]

Specimen
Serum.

Principle
The calcium in the sample is precipitated as calcium oxalate by treating with ammonium oxalate. Excess ammonium oxalate is washed off with dilute ammonium hydroxide. The calcium oxalate precipitate is then **dissolved in 1 N sulfuric acid to form oxalic acid**. It is then titrated with standard permanganate solution to find out the amount of oxalic acid which is equated with cacium present in the serum. The end point is indicated by pink color. The titer value is used to calculate the concentration of calcium.

Reagents
- 4% ammonium oxalate
- 2% ammonium hydroxide
- 1 N Sulfuric acid
- $KMnO_4$ 0.01 N–Prepare freshly before use by diluting stock 0.1 solution.

Procedure

- Take 2 mL serum in centrifuge tube.
- Add 2 mL distilled water and 1 mL of 4% ammonium oxalate to it. Mix well to achieve complete precipitation in half an hour.
- Centrifuge at 3000 rpm for 15 minutes.
- Discard the supernatant fluid without disturbing the precipitate. Invert the tubes over a filter paper to drain off the remaining supernatant for 5 minutes.
- Add 3 mL of 2% ammonia down the sides of the tube and mix the precipitate in it with a glass rod.
- Centrifuge again and pour off the supernatant.
- Add 2 mL of 1 N sulfuric acid and dissolve the precipitate in it using the glass rod used previously.
- Warm by placing it in a water bath at 70–75°C.
- Titrate it against 0.01 N permanganate taken in a microburette graduated to 0.02 mL, to get pink color that persist for a minute (gives the titer value of test).
- Perform blank titration of 0.01 N permanganate taken in a microburette against 2 mL of 1 N sulfuric acid taken in a dry test tube to the same end point as a blank (gives the titer value of blank).
- The difference between these titrations gives the volume of 0.01 N permanganate required to titrate the calcium oxalate precipitate.

Calculation
1 mL of 0.01 N permanganate is equivalent to 0.2 mg of calcium

Here 2 mL serum is used.

Mg of calcium per 100 mL of serum
$$= (\text{Titer value of test} - \text{Titer value of blank}) \times 0.2 \times 100/2$$
$$= (\text{Titer value of test} - \text{Titer value of blank}) \times 10$$

Interpretation
Reference range in adults: Total calcium– 8.6–10.2 mg% (2.15–2.55 mmol/L) Free calcium–4.6–5.3 mg% (1.15–1.33 mmol/L) (Calcium in mg% × 0.25 (conversion factor) = Calcium in mmol/L).

Alterations in serum calcium levels: The level of serum calcium is affected by:
- Defective absorption from the intestine
- Altered parathyroid hormones
- Changes in serum phosphorous concentration
- Changes in serum protein concentration
- Altered pH.

Hypocalcemia: Denotes low serum calcium levels.

Causes
- **Alkalosis:** At higher pH of alkalosis, most of the proteins acquire more negative charge which cause binding of more free calcium to proteins resulting in lowering of functional free calcium. It may manifest as tetany. Here most often total calcium remains normal but the free calcium is decreased.
- **Hypoparathyroidism:** It occurs more commonly after thyroidectomy due to the removal of a considerable part of parathyroid tissue along with thyroid gland.
- **Rickets:** Due to defective calcium absorption. Associated finding in rickets is low phosphorous levels.
- **Steatorrheas:** Due to different causes like idiopathic, celiac disease, sprue causing defective absorption leading to steatorrhea will cause low serum calcium levels.

Hypercalcemia: Denotes high serum calcium levels. Hypercalcemia occurs due to excessive release of calcium from skeleton, intestine or kidney into extracellular fluid compartment. Different conditions leading to hypercalcemia are given below:
- Primary hyperparathyroidism
- Malignancies
- Hyperthyroidism
- Hypothyroidism
- Acute adrenal insufficiency
- Renal failure
- Immobilization
- Increased serum proteins
- Lithium therapy
- Overdose of vitamin D and vitamin A.

Points to Ponder

Several calculations are used to adjust total calcium determinations to correct for variations in protein concentration. One such calculation is given below.

Corrected total calcium in mg% = Total calcium mg% + 0.8 (4-Albumin in g%): But the factors like effects of pH, binding kinetics, fatty acids, other substances bound by albumin, heparin and anions like citrate, bicarbonate, lactate, phosphate, pyruvate, β hydroxy butyrate, sulfate limit the usefulness of total and corrected calcium. So, if possible, instead of going for mathematical corrections for protein variations, it is better to determine free calcium directly by ion-selective electrode (ISE).

QUESTIONS

1. Mention two methods for estimating calcium in the serum. Give the principle of each method.
2. What are the different functions offered by calcium in the body?
3. Name the three different pools in the blood in which calcium is distributed. Give main factors affecting the redistribution in different pools.
4. What is relationship between serum proteins and serum calcium level?
5. Alkalosis causes tetany. Explain.
6. What are the anions that bind with calcium?
7. What are the normal total and free calcium levels in the serum?
8. What are the major causes for hypocalcemia?
9. What are the different causes of hypercalcemia?
10. Name the direct method of assaying free calcium.

11. Name the reference method of total calcium estimation.

REAGENT PREPARATION

All reagents should be of analytical reagent (AR) grade

- **Hydrochloric acid (0.5 mol/L):** Add 45 mL concentrated HCl to 100-200 mL distilled water in a 1 L flask and dilute up to 1 L. Thoroughly cleaned glassware are to be soaked in 0.5 M HCl overnight to remove of all traces of calcium. Then rinse with distilled water or deionized water and dry before use.
- **Buffer:** Diethanolamine 2 mol/L (pH 11.7); Urea 5 mol/L: Dissolve 210 g diethanolamine and 300 g of urea in 500 mL of distilled water. Adjust pH to 11.7 with acetic acid and dilute to 1 L. Stable for 2 months.
- **Chromogen:** Ortho-cresolphthalein complexone, 100 µmol/L; 8–Hydroxyquinoline, 8 mmol/L ; Urea 5 mol/L; Acetic acid 43mmol/L and ethanol 25% (v/v).

  Dissolve 64 mg of CPC, 1.16 g of 8–hydroxyquinoline and 2.5 mL of acetic acid in 250 mL ethanol in a 1 L volumetric flask. Add 300 g urea and dilute to 1 L with deionized water. Stable for 3 months.
- **Color reagent:** Mix equal volumes of buffer and chromogen reagent. Stable for 24 hours only.
- **Calcium standard, 2.5 mmol/L:** Dissolve 125.1 mg $CaCO_3$ (reagent grade) in 500 mL of trichloracetic acid (0.1 mol/L).

REFERENCES

1. Lorentz K. Improved determination of serum calcium with 2-cresolphthalein complexone. Clin Chim Acta. 1982;126:327-34.
2. Clark EP, Collip JB. J Biol Chem. 1925;63:461.

CHAPTER

22

Determination of Phosphate

INTRODUCTION

Phosphorous is found in the form of organic and inorganic phosphates in the body. About 85% of the total phosphate in the body is concentrated in the skeleton and the rest 15% in soft tissues. In soft tissues, most of the organic phosphates are intracellular and incorporated with proteins, nucleic acids, phospholipids and nucleotides.

The plasma contains inorganic phosphate. Serum phosphate levels are influenced by diet and parathyroid hormone (PTH) levels.

The serum phosphate is existing in three different pools (Fig. 22.1):
- 10% protein bound
- 35% complexed with sodium, calcium and magnesium

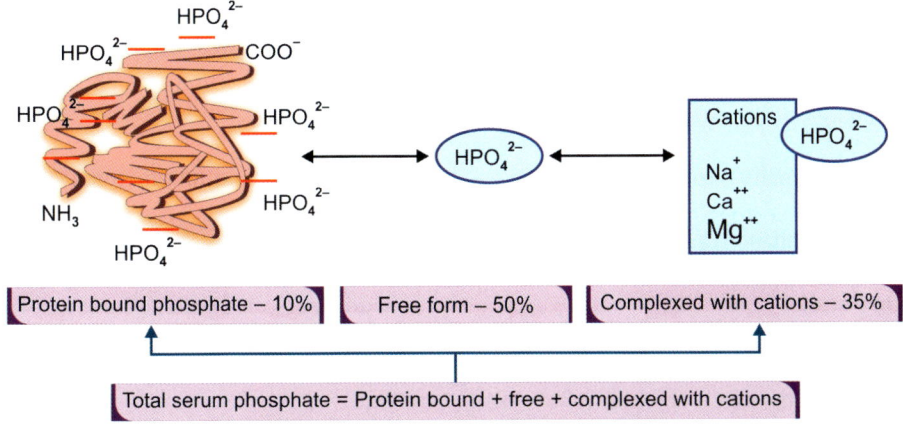

Figure 22.1 Three plasma pools of phosphate

- The rest (55%) is in free form.

    Organic phosphate esters are present inside the cellular elements of the blood.

    Apart from forming part of structural frame work of the body, the phosphate has many other **biochemical and physiological functions** in the body.
- Forming high energy phosphate bonds
- Constituent of cyclic adenine and guanine nucleotides and NADP
- Essential element of biomembranes (as part of phospholipids), nucleic acids, phosphoproteins (casein and some other proteins)
- In gene transcription and cell growth
- Regulation of enzyme action (covalent modification of enzymes by phosphorylation and dephosphorylation).

Methods

- **Chemical methods**
    - Using reaction of phosphate ions with ammonium molybdate
    - Vandate–molybdate method.
- **Enzymatic methods**
    - Monitoring the formation NADPH by enzymes (e.g. glucose 6 phosphate dehydrogenase)
    - Monitoring the formation of H_2O_2 by enzymes (e.g. xanthine oxidase)
    - Monitoring the formation NADH by enzymes (e.g. sucrose phosphorylase).

Determination of Inorganic Phosphate by Photometric Method: Method of Fiske and Subbarow[1]

Specimen
Serum is preferred.

Points to Ponder
- Serum or plasma should be separated soon after blood collection. Ester phosphates present in the cells hydrolyze to form inorganic phosphates, which diffuse out of the cell causing an elevation of its level in the serum or plasma.
- Serum phosphate levels are influenced by diet and diurnal variations (higher in afternoon and evening). To overcome this fasting specimen is better.

Principle
Serum is treated with trichloracetic acid to get protein free filtrate. Protein free filtrate is then treated with acid ammonium molybdate to form phosphomolybdic acid. The hexavalent molybdenum of phosphomolybdic acid is reduced by 1, 2, 4 amino-naphthol-sulfonic acid (ANSA) to give a blue compound, absorbance of read at 680 nm in a spectrophotometer or using red filter in a colorimeter.

Reagents Required
- Trichloroacetic acid (10%)
- Sulfuric acid 10 N
- Molybdate I
- Molybdate II
- 1,2,4 amino-naphthol-sulfonic acid (ANSA) 0.25 %
- Working standard solution (0.008 mg/mL).

Procedure
Preparation of protein free filtrate: take 2 mL serum in a dry test tube and add 8 mL TCA. Mix well and keep for 5 minutes and filter to obtain a protein free filtrate. Use this protein free filtrate for phosphate estimation (Fig. 22.2).

Calculation
5 mL of protein free filtrate contains to 1 mL serum

Mg of inorganic phosphate per 100 mL serum

Reading of test (T) = Reading of test–Reading of blank

Reading of Std (S) = Reading of std–Reading of blank

Take three '10 mL' stoppered test tubes and mark T, S and B **To 'T' tube** : Add PFF – 5 mL, Molybdate II – 1 mL. Mix well and add ANSA 0.4 mL and make up to 10 mL mark with distilled H$_2$O **To 'S' tube** : Add Std – 5 mL, Molybdate I – 1 mL. Mix well and add ANSA 0.4 mL and make up to 10 mL mark with distilled H$_2$O **To 'B' tube** : Add water – 5 mL, Molybdate II – 1 mL. Mix well and add ANSA 0.4 mL and make up to 10 mL mark with distilled H$_2$O		
T: PFF – 5 mL, Molybdate II – 1 mL, Mixwell, ANSA – 0.4 mL, Make up to 10 mL by water	S: Standard – 5 mL, Molybdate I – 1 mL, Mixwell, ANSA – 0.4 mL, Make up to 10 mL by distilled water	B: Distilled water – 5 mL, Molybdate II – 1 mL, Mixwell, ANSA – 0.4 mL, Make up to 10 mL by distilled water
Mix gently the contents of the tube after each addition and keep for 5 minutes Read at 680 nm in a spectrophotometer or in a colorimeter using red filter		

Figure 22.2 Procedure of method of Fiske and Subbarow[1] for inorganic phosphate estimation

Concentration of standard in 5 mL of working std solution = 0.008 × 5 = 0.04 mg

= T/S × Concentration of standard × 100/ Volume of serum

= T/S × 0.04 × 100/1 = **T/S × 4 mg%**

(Phosphate in mg% × 0.323 (conversion factor) = Phosphate in mmol/ L).

Interpretation
Reference range of serum inorganic phosphate
Adults: 2.5–4.5 mg%   (0.81–1.45 mmol/L)
Children: 4.0–7.0 mg%   (1.29–2.26 mmol/L)

Alterations in inorganic phosphate in serum.

Hypophosphatemia

The term hypophosphatemia is used when the serum inorganic phosphate concentration is less than 2.5 mg%. Clinical manifestations depend on duration and extent of the deficiency. Since phosphate is a component of energy currency of the—ATP, cellular functions are impaired in hypophosphatemia.

It leads to muscle weakness, respiratory failure decreased cardiac output. At very low concentrations like, below 0.5 mg%, rhabdomyolysis, hemolysis, mental confusion and even coma may occur. Chronic hypophosphatemia causes rickets in children and osteomalacia in adults.

Causes

- **Oral or intravenous hyperalimentation and insulin:** Carbohydrates induce insulin secretion which enhances transport of phosphate from extracellular fluid into the cells leading to a fall in serum phosphate.
- **Respiratory alkalosis:** Promote an intracellular shift of phosphate from extracellular fluid leading to a fall in its level in the serum.
- **Lowered renal threshold for phosphate**
  - Primary and secondary hyperparathyroidism
  - Fanconi's syndrome
  - X linked hypophosphatemia.

- **Intestinal loss**
  - Malabsorption
  - Ingestion of antacids containing aluminum and magnesium which bind with phosphate making it unsuitable for absorption.
- **Acidosis**, e.g. ketoacidosis, lactic acidosis. Acidosis leads to catabolism of organic phosphates to form inorganic phosphates which then pass into plasma and excreted in urine leading to depletion of intracellular phosphates.

Hyperphosphatemia

Elevated phosphate causes a decrease in serum calcium concentration which may lead to tetany and seizures. An increase in serum phosphorous is seen in the following conditions:
- Renal failure: A decrease in glomerular filtration rate decrease phosphate excretion in urine leading to hyperphosphatemia
- Hypoparathyroidism and acromegaly: Enhances tubular reabsorption of phosphates
- Aggressive phosphate therapy
- Release of phosphate: Occurs in rhabdomyolysis, chemotherapy.

QUESTIONS

1. How will you estimate inorganic phosphate in the serum? Name one method, give its principle.
2. What will happen to phosphate value if plasma or serum separation of blood sample is delayed? Give the reason.
3. Name the 3 different pools in the blood in which serum phosphates are distributed.
4. Give the reference range of serum phosphate in adults and children.
5. What is hypophosphatemia? Name different conditions leading to hypophosphatemia.
6. What is hyperphosphatemia? Name different conditions leading to hyperphosphatemia.
7. Why fasting serum specimen is considered to be ideal for phosphate estimation?
8. What are the manifestations of hyperphosphatemia?
9. What are the manifestations of hypophosphatemia?
10. Why do you get hypophosphatemia in respiratory alkalosis?
11. Renal failure cause hyperphosphatemia. Explain.

REAGENT PREPARATION

- **Trichloroacetic acid (TCA) 10% (w/v):** Dissolve 10 g of reagent grade TCA in water and make up to 100 mL in a 100 mL conical flask or cylinder.
- **Sulfuric acid 10 N:** Add 450 mL concentrated sulfuric acid **slowly** into 1300 mL of distilled water.
- **Molybdic acid reagent I:** (2.5% ammonium molybdate in 5 N sulfuric acid.) Dissolve 25 g ammonium molybdate in 200 mL distilled water and transfer it to a 1 L flask containing 500 mL 10 N sulfuric acid. Dilute to 1 L with water.
- **Molybdic acid reagent II:** (2.5% ammonium molybdate in 3 N sulfuric acid). Dissolve 25 g ammonium molybdate in 200 mL distilled water and transfer it to a 1 L flask containing 300 mL 10 N sulfuric acid. Dilute to 1 L with water.
- **Sodium bisulfite 15% solution:** Dissolve 15 g of sodium sulfate in 100 mL distilled water. Freshly prepared solutions may be turbid so keep it for 2 days to make it clear.
- **Sodium sulfite 20% solution:** Dissolve 200 g sodium sulfite ($Na_2SO_3 7H_2O$) in 380 mL water. Filter and keep it in a stoppered bottle.

- **0.25% 1,2,4 amino-naphthol-sulfonic acid:** Add 0.5 g to 195 mL of 15% sodium bisulfite and 5 mL of 20% sodium sulfite. Stopper and shake well to dissolve it. Store in an amber colored bottle in cold. Solution is stable up to 4 weeks.
- **Standard phosphate solution (0.4 mg/ 5 mL):** Dissolve 0.351 g of pure potassium dihydrogen phosphate in water in a liter flask add 10 mL of 10 N sulfuric acid and make up to the mark with water. Five mL contains 0.4 mg. This is stable.
- **Working standard (0.008 mg/mL):** Dilute the stock standard 1 to 10 so that 1 mL contains 0.008 mg phosphorous.

REFERENCE

1. Fiske CH, Subbarow Y. The colorimetric determination of phosphorous. J Biol Chem. 1925;66:375-400.

SECTION 3

Charts

CHAPTER 23

Charts

ACID BASE DISORDERS

Reference ranges of acid base parameters

pH	7.35–7.45
pCO_2	35–45 mm Hg
HCO_3^-	22–26 mmol/L
$cdCO_2$	1.07–1.38 mEq/L
$HCO_3^-/cdCO_2$	20:1

More about $cdCO_2$: $cdCO_2$ (concentration of dissolved carbon dioxide) represent both undissociated CO_2 (H_2CO_3) and free CO_2 (physically dissolved). At the physiological pH of blood, dissolved CO_2 fraction is about 700–1000 times greater than H_2CO_3. Hence, $cdCO_2$ used to represent combined concentration of undissociated CO_2 (H_2CO_3) and free CO_2 (physically dissolved).

$cdCO_2$ is calculated by multiplying pCO_2 in mm Hg with solubility coefficient of CO_2 in blood at 37°C [solubility coefficient (α) = 0.0306]. [i.e. when pCO_2 = 40 mm Hg, $cdCO_2$ = 40 × 0.0306 = 1.224 mmol/L].

Q. 1

Sheela aged 19 years brought to the causality at 11 AM with dizziness, tingling of fingers, sweating and breathing deeply and nausea. On examination: Hyperventilation, carpopedal spasm were found.

Laboratory data:

pH	pCO_2	HCO_3^-	$cdCO_2$
7.55	28 mm Hg	22 mmol/L	0.86 mEq/L

- What kind of acid base disorder, this girl is suffering from? Explain.
- What are the common causes of this kind of acid base disorder?
- Give the compensatory mechanisms available in the body to correct this sort of acid-base imbalance?

Ans. 1

- Respiratory alkalosis (uncompensated)
  In this case, pH >7.45, which indicates alkalosis (pH range in health: 7.35–7.45).
  To find the type of alkalosis
  – Here, the metabolic parameter HCO_3^- is 22 mmol/L, which is within reference range of 22–26 mmol/L, hence, it is not metabolic alkalosis.
  – But the respiratory parameter pCO_2 is 28 mm Hg. It is lower than reference

range of 35–45 mm Hg indicating involvement of respiratory system. Here there is CO_2 washout due to hysterical breathing → ↓pCO_2.

∴ It is respiratory alkalosis.

To find out whether it is compensated or uncompensated

Whenever there is respiratory alkalosis, renal acid base regulatory mechanisms become active, leading to decreased reclamation of HCO_3^-, resulting in ↓ serum of HCO_3^-. Normal serum HCO_3^- in this case indicates that the stage of compensation has not occurred.

Henderson-Hasselbalch equation is useful in this context as shown below:

pH = pKa + log [HCO_3^-]/[$cdCO_2$]

pka of bicarbonate/carbonic acid buffer system = 6.1

[$cdCO_2$] = pCO_2 × 0.0306 (solubility coefficient of CO_2 in blood

Normally the average ratio of serum bicarbonate to carbonic acid [$cdCO_2$] is 20.

At the normal pH of blood, a pattern of Henderson-Hasselbalch equation is shown below as an example, to show the normal ratio of serum bicarbonate to carbonic acid (20).

7.4 = 6.1 + log 25/1.25
7.4 = 6.1 + log 20

In this case, [HCO_3^-]/[$cdCO_2$] = 22/28 × 0.0306 = 22/0.86 = 25.6.

The ratio is >20, with high pH.

∴ *It is uncompensated respiratory alkalosis.*

Tingling of fingers and carpopedal spasm is due to a reduction in ionized calcium caused by increased binding of calcium to proteins in alkaline pH of blood.

- Common causes of respiratory alkalosis:
  – Functional hyperventilation in hysteria and anxiety disorders
  – Cases where the respiratory center in the medulla is over stimulated as in encephalitis, intracranial surgery, salicylate poisoning, high altitude (mountain climbers) and chronic liver disease.
- **The compensatory mechanisms**
  – **At the level of buffers**: Buffers in tissues and blood provide H^+ ions to bind with HCO_3^- as shown below. But this is a minor mechanism.
  $HCO_3^- + H^+ \rightarrow H_2CO_3 \rightarrow H_2O + CO_2$
  – **At the level of kidneys**: The following mechanisms will help to reduce serum HCO_3^- level → lowering of HCO_3^-/H_2CO_3 ratio to 20 and pH back to normal.
    - ↓ Na^+ – H^+ exchange
    - ↓ formation of ammonia and ammonium ions
    - ↓ reclamation of HCO_3^-

Q. 2

Simon 75 years who has undergone an intracranial surgery showed following pattern of acid-base parameters. Interpret it.

Laboratory data:

pH	pCO_2	HCO_3^-	$cdCO_2$
7.5	24 mm Hg	20 mmol/L	0.73 mEq/L

Ans. 2

- **Respiratory alkalosis**
  Here pH is >7.45 indicating alkalosis (normal pH range =7.35–7.45)
  pCO_2 = 24 mm Hg which is less than normal (normal pCO_2 = 35–45 mm Hg)
  HCO_3^- = 20 mEq/L (normal HCO_3^- = 22–26 mmol/L)
  [HCO_3^-]/[$cdCO_2$] = 27.4 (20/24 × 0.0306 = 20/0.73 = 27.4) more than 20. Uncompensated type of alkalosis.
  ∴ *Uncompensated respiratory alkalosis* (For details see Ans. 1).
  The probable reason in this case is the intracranial surgery causing respiratory center stimulation leading to wash out of CO_2 resulting in ↓pCO_2, ↑pH and HCO_3^-/$cdCO_2$ ratio.

Q. 3

Ramakrishnan, 60 years old a known smoker attended causality with exacerbation of bronchial asthma. The acid base analysis report is given below. Give your interpretation.

Laboratory data:

pH	pCO$_2$	HCO$_3^-$	cdCO$_2$
7.04	90 mm Hg	24 mmol/L	2.7 mEq/L

Ans. 3

Respiratory acidosis (Uncompensated)
Here pH is <7.35 indicating acidosis (normal pH range = 7.35–7.45).

Type of acidosis: In this case metabolic parameter HCO$_3^-$ is within reference range of 22–26 mmol/L, hence excludes metabolic acidosis.
But the respiratory parameter pCO$_2$, is 90 mm Hg which is higher than reference range of 35–45 mm Hg indicating CO$_2$ retention due to asthma.
∴ *It is respiratory acidosis.*

Asthma attacks are characterized by episodes of airway obstruction causing CO$_2$ retention → ↑ pCO$_2$, ↓ pH, ↓ bicarbonate – carbonic acid ratio [HCO$_3^-$]/[cdCO$_2$] and normal bicarbonate in the initial phase.

To find out whether it is compensated or uncompensated
In this case, the normal serum HCO$_3^-$ suggests that the stage of compensation has not been achieved.
[HCO$_3^-$]/[cdCO$_2$] = 24/90 × 0.0306 = 24/2.8 = 8.6

The ratio of bicarbonate to carbonic acid [HCO$_3^-$]/[cdCO$_2$] is 8.6:1, indicating bicarbonates are not reclaimed and CO$_2$ not eliminated sufficiently to raise ratio to normal 20: 1
∴ *It is uncompensated respiratory acidosis.*

Compensatory mechanisms of respiratory acidosis

At the level of buffers: Excess carbonic acid produced by retention of CO$_2$ is buffered by Hb and proteins to certain extent.

At the level of lungs: Raised pCO$_2$ causes stimulation of respiratory center (in cases where there is no involvement of respiratory center) → ↑rate and depth of breathing → CO$_2$ elimination through lungs → ↓cdCO$_2$. These changes will help to raise the ratio [HCO$_3^-$]/[cdCO$_2$] to certain extent. But it is difficult in a case of obstructive airway disease.

At the level of kidneys: The following mechanisms will help to increase the serum HCO$_3^-$ level and bring back the HCO$_3^-$/H$_2$CO$_3$ ratio to 20.
- ↑ Na$^+$ – H$^+$ exchange
- ↑ formation of ammonia and ammonium ions
- ↑ reclamation of HCO$_3^-$.

Causes of respiratory acidosis: Generally respiratory acidosis is caused by:
- Disorders that interfere with the respiratory activity, e.g. Pneumonia, asthma, pulmonary edema, emphysema, apnea.
- Depression of respiratory center, e.g. Morphine and barbiturate poisoning.

Q. 4

A woman complaining of intractable vomiting, suspected of pyloric stenosis and receiving treatment showed following acid base data on day 1 and day 2. Comment on:

Laboratory data on day 1 and day 2

Day	pH	pCO$_2$	HCO$_3^-$	cdCO$_2$
Day 1	7.6	40 mm Hg	35 mmol/L	1.2 mEq/L
Day 2	7.55	45 mm Hg	28 mEq/L	1.38 mEq/L

Ans. 4

Day 1: Metabolic alkalosis (uncompensated)
It is a case of alkalosis, because pH is >7.45.

Type of alkalosis: Metabolic alkalosis.

Reason: Respiratory component (pCO_2) is normal but the metabolic component HCO_3^- is raised.

Level of compensation: pH is high and $HCO_3^-/cdCO_2$ is >20 (here it is 29)
∴ *Uncompensated metabolic alkalosis.*

Day: 2 Metabolic alkalosis (partially compensated).
Here pH is >7.45, indicating alkalosis.

Type of alkalosis: Metabolic alkalosis.

Reason: Respiratory component (pCO_2) is normal but the metabolic component HCO_3^- is raised.

Level of compensation: pH is high and the ratio brought to 20 ($HCO_3^-/cdCO_2 = 20$). It has reached a partially compensated stage.
∴ *Partially compensated metabolic alkalosis*

Compensatory mechanisms that occur during metabolic alkalosis

At the level of lungs: Alkaline pH depress the respiratory center →↓ respiratory rate and depth → retention of CO_2 → ↑ H_2CO_3 and ↑ $cdCO_2$.

At the level of kidneys:
- ↓ $Na^+ - H^+$ exchange
- ↓ formation of ammonia and ammonium ions
- ↓ reclamation of HCO_3^-

Thus, respiratory and renal mechanisms have resulted in changes, which are evident as partially compensated metabolic alkalosis on day 2.

Causes of metabolic alkalosis
It can be due to either excess of base or loss of acid.
- Loss of acid
  - From the stomach—severe vomiting, aspiration of gastric contents
  - Through urine—diuretic drug therapy (carbonic anhydrase inhibitors and potassium sparing drugs).
- Endocrine disorders, e.g. Primary aldosteronism and Cushing syndrome cause hypokalemic alkalosis (retention of sodium and loss of potassium → hypokalemia which in turn causes shift of protons into the ICF → alkalosis).

Q. 5

Kumaran, 58 years old, peon in a private firm, has been suffering from diabetes mellitus for the past 20 years. He was taking irregular treatment for DM. On one Sunday he was brought to the casuality in a stuporous state. On examination, stuporous, fruity smell +, Kussmaul's type of breathing +
Urine—Rothera's test +ve and Blood glucose—450 mg%. Acid base data of his blood is given below. Comment on the acid base status of the patient

Laboratory data:

pH	pCO_2	HCO_3^-	$cdCO_2$	Na^+	K^+	Cl^-
7.2	40 mm Hg	15 mmol/L	1.2 mEq/L	140 mmol/L	4 mmol/L	102 mmol/L

Ans. 5

Uncompensated metabolic acidosis (Diabetic ketoacidosis).

Acidosis: Because pH is 7.2, which is less than normal pH (7.35–7.45).

Type of acidosis: Metabolic acidosis.

Reason: Respiratory component (pCO_2) is normal but the metabolic component HCO_3^- is reduced.

Level of compensation: pH is low and the bicarbonate – carbonic acid ratio is <20 (HCO_3^-/ $cdCO_2$ = 15/1.2 = 12.5). These values indicate uncompensated state.
∴ *Uncompensated metabolic acidosis.*

About DKA (Diabetic ketoacidosis): In uncontrolled diabetes mellitus due to altered insulin: glucagon ratio (i.e. deficient action of insulin and over action of glucagon) ketoacids (acetoacetate, beta hydroxybutyric acid and acetone) are produced excessively which are acidic in nature. These ketoacids ionize to release protons and respective anions as shown below. Release of protons → acidemia

　Acetoacetic acid → Acetoacetate$^-$ + H$^+$
　β Hydroxybutyric acid → β Hydroxybutyrate + H^1
　Acetone being volatile, is mainly excreted in expired air.

Compensatory mechanisms in metabolic acidosis

At the level of buffers: Bicarbonate–carbonic acid buffer tend to neutralize excess protons generated during acidosis and HCO_3^- used up in this process causing lowering of bicarbonate level.

$$HCO_3^- + H^+ \rightarrow H_2CO_3 \rightarrow CO_2 + H_2O$$

At the level of lungs: Acidic pH stimulate respiratory center →↑ respiratory rate and depth of respiration → elimination of H_2CO_3 as CO_2 → ↓pCO_2 resulting in decrease in $cdCO_2$. The presence of the volatile ketone body in the expired air will impart a fruity odor to the breath (Kussmaul's respiration).

At the level of kidneys: Renal mechanisms tend to restore the normal pH by increasing the excretion of acid in urine by the following mechanisms:
- ↑ Na$^+$ – H$^+$ exchange
- ↑ formation of ammonia and ammonium ions
- ↑ reclamation of HCO_3^-.

Types of metabolic acidosis: Normally the sum of anions is equal to the sum of cations. But certain anions like proteins, sulfates and phosphates cannot be measured. There is a difference between measured cations and anions which is described as anion gap which constitutes unmeasured anions.

　Anion gap = [Na$^+$ + K$^+$] – [HCO_3^- + Cl$^-$]

The apparent gap is due to proteins, sulfates and phosphates in a healthy person. The **normal anion gap is 7–16 mmol/L.**

In this case, the anion gap is 27 (AG in this case = (140 + 4)-(102 + 15) = 27). Therefore, it is a case of **high anion gap acidosis** due to abnormal accumulation of anions (ketone bodies).

Different types of metabolic acidosis can be differentiated on the basis of anion gap.
- **High anion gap acidosis**
  Ketoacidosis—starvation, diabetes mellitus
  Renal failure → uremia
  Lactic acidosis
  Salicylate toxicity
- **Normal anion gap acidosis**
  Gastrointestinal fluid loss – diarrhea
  Renal tubular acidosis.

JAUNDICE

Reference ranges (Normal pattern) of Liver Function Tests required for the diagnosis of Jaundice.

Blood

Total bilirubin (TB)	0.3–1.2 mg%
Conjugated bilirubin (CB)	0–0.2 mg%
Unconjugated bilirubin (UCB)	= TB–CB

Serum Enzymes

ALT (Alanine amino transferase) (Older terminology – SGPT)	13 – 40 U/L
AST (Aspartate amino transferase) (Older terminology –SGOT)	8 – 20 U/L
ALP (Alkaline phosphatase)	40–125 U/L
GGT (Gamma-glutamyl transferase)	<55 IU/L
Albumin	3.5–5.0 g%
A:G ratio	1.2–2.5: 1

Urine

Bile pigment: (Modified Fouchet's test)	Negative
Bile salt: (Hay's Test)	Negative
Urobilinogen: (Ehrlich's test)	Faintly positive

Q. 6
Resmi, 18 years staying in a college hostel, brought to the outpatient clinic of Medical College with complaints of fever, headache, nausea and yellowish discoloration of sclera On examination, Febrile, jaundice + Liver palpable.

Laboratory data:

Blood	TB	CB	ALT	AST	ALP
	7.2 mg%	2.6 mg%	160 IU/L	70 IU/L	180 IU/L
Urine	Bile salts + ve	Bile pigments +ve	Urobilinogen Trace		

What is the probable diagnosis? Evaluate the clinical condition by the laboratory data provided.

Ans. 6
Probable diagnosis: Hepatic jaundice.

The **clinical features** fever, headache and nausea are suggestive of an infection and the finding of liver enlargement with yellowish discoloration of sclera suggests hepatic jaundice.

Laboratory data confirms the hepatic origin of jaundice.

Serum bilirubin levels:
- Elevated **TB** levels suggest jaundice. *Jaundice clinically manifest at bilirubin levels ≥ 3 mg%.* Hepatocyte dysfunction due to hepatitis →↓ glucoronyl transferase activity → elevation of unconjugated bilirubin **(UCB)** (UCB = 6-2.6 = 3.4 mg%).
- **CB** is also raised due to blockade of biliary microchannels by inflammation leading to regurgitation of biliary contents into blood.

Serum enzymes:
- Rise in **aminotransferases** (ALT and AST) shows inflammatory injury to hepatocyte causing leaking of cytoplasmic enzymes (amino transferases) of hepatocytes into circulation.
- Slight elevation of **alkaline phosphatase** points towards the release of membrane bound ALP into blood, resulting from the back pressure produced by obstructed inflamed biliary channels.
- **Urinary findings** of **positive bile salts** and **bile pigments** again indicate that the patient is in the obstructive phase of hepatic jaundice (i.e. infection causing inflammation of lining cells of biliary canaliculi causing regurgitation of biliary contents into the bloodstream. When the blood levels of these compounds crosses the renal threshold for that substances, they are excreted in urine, thus CB and bile salts are excreted in urine).
- Urobilinogen is in trace excludes the possibility of hemolytic jaundice.

How to differentiate hepatic jaundice from obstructive jaundice due to stones, tumors or other obstructions in the biliary tract ?
- **Serum bilirubin values:** In obstructive jaundice, the level of CB will be much higher than the hepatic jaundice and UCB values remain within normal limits.

- **Enzymes:** Transaminase values generally remain within normal range but ALP values will be very high.
- **Urinary findings:** Bile salts and CB will be positive but urobilinogen will be absent (Due to biliary obstruction CB cannot reach the intestine in obstructive jaundice and hence urobilinogen cannot be formed). Urine will give a negative response to Ehrlich's test (test for urobilinogen) and the patient will complain of passing clay colored stools (due to the absence of stercobilinogen in feces) in obstructive jaundice.

Q. 7

Kurinji, 45 years old woman, a tribal hailing from Waynaud district came with severe tiredness and severe pain all over the body. On examination, Pallor + jaundice + hepatosplenomegaly. Based on clinical and laboratory data what is your provisional diagnosis? What other test you require to confirm the diagnosis?

Laboratory data	Reference range
Blood	
Hb	7 g%
Reticulocytes	10%
Sickling test	+ve
Total bilirubin (TB)	10 mg% (0.3–1.2 mg/dL)
Conjugated bilirubin (CB)	0.6 mg% (0.1–0.4 mg/dL)
Unconjugated bilirubin	9.4 mg% (0.2–0.7 mg/dL)
ALP	45 IU/L (40–125 IU/L)
ALT (SGPT)	14 IU/L (10–35 IU/L)
AST (SGOT)	20 IU/L (8–20 IU/L)
Urine	
Bile salts	Negative
Bile pigment	Negative
Urobilinogen	Strongly positive

Ans. 7

Provisional diagnosis: The woman is suffering from hemolytic jaundice probably due to sickle cell disease.

- Total bilirubin is high (10.0 mg%) → suggest jaundice
- UCB = 10.0 − 0.6 = 9.4 mg% is also high
- CB = within normal limits

CB and UCB values indicate prehepatic jaundice here the underlying problem is sickle cell disease → exaggerated hemolysis → ↑ TB and ↑UCB. Excess bilirubin formed saturate the bilirubin conjugation capacity of the liver →↑UCB. This finding excludes obstructive jaundice.

∴ ↑ *TB and UCB indicate a hemolytic disorder.* (Here hemolytic disorder is caused by sickle cell disease).

- Serum enzyme studies show normal pattern which indicates hepatocytes are not involved in the disease process thereby excluding hepatic jaundice and obstructive jaundice and confirms diagnosis of hemolytic jaundice.
- Urine—Bile salts negative which shows that the jaundice is not obstructive type.
- Urobilinogen strongly positive: It is due to increased rate of RBC breakdown →↑ heme release and catabolism →↑CB formation and secretion in bile → more CB reaches intestine. In the intestine, the excessive amount of bilirubin converted into urobilinogen and stercobilinogen. More than normal amount of urobilinogen is absorbed from the intestine into the blood and increasing amounts excreted in urine.
- The positive sickling test, tribal origin of the woman and the kind of pain suggestive of sickling crisis strongly suggests sickle cell disease.

Sickle cell disease can be confirmed by:
- Hb electrophoresis which show the presence of Hb S
- Globin chain separation to demonstrate abnormal 'S' globin chain.

Q. 8

Meenakshi, a 58-year-old woman case of pain in the upper right side of the abdomen, fever with chills, pruritus and passing dark urine and clay colored stools. On examination, Jaundice + scratch marks on the skin + Fever. No other findings obtained. Her laboratory report is shown below.

Laboratory data:

Blood	TB	CB	ALT	AST	ALP
	12.0 mg%	10 mg%	30 IU/L	18 IU/L	200 IU/L
Urine	Bile salts +ve	Bile pigments +ve	Urobilinogen -ve		

What is your diagnosis? Explain.

Ans. 8

The lady is suffering from obstructive jaundice (cholestasis)
- High TB → suggest jaundice.

To find out the type of jaundice
- CB is high which suggest, obstruction in the biliary passages leading to cholestasis → regurgitation of CB into blood →↑CB

- Serum enzyme studies show **high ALP** indicating obstructive type of jaundice and **normal transaminases levels** show that hepatocytes are unaffected by any disease process thereby excluding hepatic jaundice.
- **Urine:** Positive modified Fouchet's test and positive Hay's test: supporting the diagnosis of obstructive jaundice. Obstruction of biliary passages causing stasis leading to regurgitation of bile constituents into the blood thereby elevating the concentration of CB and bile salts in the blood and their excretion in urine → Positive modified Fouchet's test and Positive Hay's test. Bile salts have a tendency to get deposited in the skin causing **intense pruritus.**
- **Test for urine urobilinogen (Ehrlich's test) is negative:** Due to obstruction in the biliary passages biliary contents cannot reach the intestine. Therefore, intestinal bacteria cannot act upon CB to form bilinogens (urobilinogen and stercobilinogen). This is the reason for negative Ehrlich's test for urobilinogen and clay colored stools.

Q. 9

Go through the following table and comment on

Serum bilirubin (mg%)	ALP (IU/L)	Aminotransferases (IU/L)	Urine bilirubin	Urine bile salt	Urine urobilinogen	Comment on the condition
TB – 8.0 CB – 3.5 UCB – 4.5	150 IU/L	ALT- 160 AST - 80	Present	Present	Trace	a) ?
TB – 7.0 CB – 0.3 UCB – 6.7	88 IU/L	ALT- 25 AST - 15	Absent	Absent	Increased	b) ?
TB – 0.8 CB – 0.3 UCB – 0.5	60 IU/L	ALT- 20 AST - 10	Absent	Absent	Trace	c) ?
TB – 14 CB – 13 UCB – 1.0	280 IU/L	ALT- 30 AST - 18	Present	Present	Absent	d) ?

Ans. 9
a) Hepatic jaundice, b) Hemolytic jaundice, c) Normal, d) Obstructive jaundice.

DIABETES MELLITUS

Diagnostic criteria for diabetes mellitus: As per American Diabetes Association (ADA)*2015 guidelines
Any one of the following is diagnostic of diabetes mellitus:
- In a patient with classic symptoms of hyperglycemia (polyuria, polydypsia and polyphagia), a random (regardless of the time of the preceding meal) plasma glucose ≥200 mg% (11.1 mmol/L).
- Fasting (no caloric intake for at least 8 hours) plasma glucose (FPG) ≥ 126 mg% (7 mmol/L)**
- 2-hour post load plasma glucose concentration ≥200 mg% (11.1 mmol/L) during an OGTT. **
- HbA_{1c} >6.5% **, ***

Other tests used in DM for risk assessment and their reference range
- Lipid profile with desirable levels
  - Triglycerides (TG): <150 mg% (<1.7 mmol/L)
  - Total cholesterol (TC): <5.2 mmol/L (<200 mg%)
  - HDL cholesterol: (HDLc) >1.0 mmol/L (>40 mg%) in males; >1.3 mmol/L (>50 mg%) in females
  - LDL cholesterol (LDLc): ≤3.5 mmol/L (<130 mg%)

---
*American Diabetes Association. Classification and diagnosis of diabetes. Sec.2. *In* Standards of Medical Care in Diabetes – 2015. Diabetes Care 2015; 38 (Suppl.1): S8-S16.
**In the absence of unequivocal hyperglycemia, results should be confirmed by repeat testing.
***HbA_{1c} testing should be performed with a method standardized to DCCT (Diabetes Control and Complications Trial) reference assay.

- Microalbuminuria: 30–300 mg/24 hours
- HbA_{1c} (Glycated hemoglobin): 4.0–5.6%.

Q. 10
A 50-year-old executive attended outpatient clinic with case of passing large volumes of urine at frequent intervals, unusual thirst and overeating and a feeling of weight loss for the past 3 months.
Interpret the results. What other tests you like to do?

Urine

Dipstick test for glucose: Positive.

Blood

Random plasma glucose on two separate occasions: 270 mg%(15 mmol/L) and 288 mg% (16 mmol/L).

Ans. 10
Diagnosis: Diabetes mellitus.

Reason:
- Classical symptoms of DM—Polyuria, polydypsia and polyphagia are present.
- Random plasma glucose values on two separate have met the diagnostic criteria of DM—more than 200 mg%.

Other Useful Tests

- HbA_{1c} will give an idea about the glucose level in the previous 3 months and also it can serve as an index of long-term glycemic control.
  During persistent elevation of plasma glucose, hemoglobin undergo nonenzymatic glycation at an increased rate. So the level of HbA_{1c} gives an idea about the glucose level over the previous 3 months (since the life span of RBC is 120 days).

Significance of HbA$_{1c}$ Levels

6.0–6.4%	Prediabetes*
>6.5%	Diabetes
≤7.0%:	Target value to be achieved in Type 1 and 2 DM **to reduce microvascular complications**

Note: 1% rise in HbA$_{1c}$ correspond to 35 mg% rise in plasma glucose.

*American Diabetes Association. Diagnosis and classification of diabetes mellitus. Diabetes Care. 2012:35(Suppl 1):S64-71.

- **Lipid profile**
  Total cholesterol, HDL-Cholesterol, LDL-cholesterol and triglycerides will help estimating the risk of developing atherosclerotic vascular disease. If the levels are not within desirable limits, lipid lowering drugs should be started.
  Desirable levels of lipid parameters: Triglycerides <150 mg%; TC - <200 mg%; HDL- C >60; LDL-C <130 mg%.
- **C-Peptide assay:** Though not routinely done for routine management of patients with diabetes, it is important in clinical research.
  - It can give an idea about the insulin secretion capacity of an individual
  - It gives an estimate of endogenous insulin secretion in patients on insulin therapy (C-peptide (connecting peptide) is produced during post-translational modification of endogenous insulin and it is not present in commercial insulin preparations of insulin).

Q. 11

Two brothers, Peter and Joseph aged 44 and 38 years came to attend a preventive health check-up camp because of strong family history of diabetes mellitus. They were examined clinically and based on preliminary investigations, they were advised to undergo oral glucose tolerance test (OGTT). The relevant details of the test report are given above:

- Give your interpretation and guidelines if any, for further follow-up
- What are the indications for OGTT?

Investigations	Peter 44 years	Joseph 38 years
Urine dipstick test for glucose	Negative	Negative
Fasting plasma glucose	118 mg% (6.6 mmol/L)	98 mg% (5.4 mmol/L)
2- hour post load plasma glucose concentration during OGTT	138 mg% (7.7 mmol/L)	150 mg% (8.3 mmol/L)

Ans. 11

Peter is suffering from impaired fasting glucose (IFG).

IFG is a stage of impaired glucose homeostasis. To diagnose IFG, the glucose values should be as shown below:

Fasting plasma glucose concentrations of 100–125 mg% (5.6–6.9 mmol/L).*

Joseph is suffering from impaired glucose tolerance (IGT).

IGT is a stage of impaired glucose regulation manifesting biochemically as glucose tolerance above the normal limits but lower than the cut off values to diagnose diabetes mellitus.

To diagnose IGT, the following criterion must be met: **2-hour post load plasma glucose concentration during OGTT should be between 140 and 199 mg% (7.8 mmol/L and 11.1 mmol/L). IGT is a transient stage between normal glucose tolerance and type II diabetes mellitus (prediabetic stage).

---

*[Forouchi NG, Balkau B, Borch Johnson K, et al. EDEG. The threshold for diagnosing impaired fasting glucose: a position statement by the European Diabetes Epidemiology Group. Diabetologia. 2006; 49:8222-7].

** [Shaw JE, Zimmer PZ, Alberti KG. Point:Impaired fasting glucose: the case for the new American Diabetes Association Criterion. Diabetes Care 2006:29:1170-2].

Significance of recognizing IFG and IGT
Both IFG and IGT are associated with increased risk of developing diabetes in future (prediabetic state). Risk is more when both coexist in one individual.

Lifestyle modifications (exercise, healthy food habits) are effective in delaying or even preventing the onset of diabetes in persons with IFG and/or IGT.

Indications for OGTT
- Diagnosis of gestational diabetes mellitus (GDM)
- IFG and IGT
- Pregnant ladies with past history of big baby, i.e. more than 4 kg or a past history of miscarriage
- To rule out benign renal glucosuria.

Test

Instructions
- Three days of unrestricted carbohydrate diet (>150 g) to sensitize the β cells of pancreas
- Test is done in the morning (between 0700 and 0900 hours) after 10–12 hours of overnight fast (intake of water is permitted)
- Smoking and physical activity should be avoided.

Procedure
- Patient should be seated comfortably
- Withdraw blood in the fasting state for testing plasma glucose level
- Give a glucose load 75 g for adults and 1.75 g/kg body weight for children (maximum of 75 g) dissolved in 300 mL of water and drunk within 5 minutes. Then every 30 minutes withdraw blood for 2 hours after the glucose load.

Urine may be collected with every blood sample collection to test the presence of glucose.

To rule out diabetes mellitus by OGTT
FPG <100 mg% and 2-hour post glucose load plasma glucose <140 mg % → rule out diabetes mellitus.

Q. 12
Mary, 42-year-old working woman have undergone a health check-up to join a health insurance scheme. A part of the investigation report is given below for evaluation. What is your impression? Is there any additional test required?

Urine

Dipstick test for glucose (fasting and post-glucose load urine specimens)—Negative.

Blood

Fasting plasma glucose—118 mg%.

Ans. 12
Fasting plasma glucose value suggest impaired fasting glucose (which is considered to be a forerunner of type 2 diabetes mellitus).

Further evaluation need study of tolerance of glucose under standard conditions, i.e. oral glucose tolerance test (OGTT).

Q. 13
Suhara 35-year-old obese (Height: 160 cm, Weight: 72 kg) pregnant lady visited Obstetrics and Gynecology OPD in the second semester for check-up. It was her second pregnancy. She told that her first baby weighed 4.3 kg and her mother died of diabetes mellitus. She had no other complaints, and on examination, also she appeared normal except for the obesity. What should be done for her?

Ans. 13
Obesity, history of having given birth to an overweight baby and family history of diabetes

mellitus are risk factors pointing towards gestational diabetes mellitus (GDM).

She should be screened for GDM.

Screening for GDM should be routinely done on all pregnant women ≥25 years or <25 years with any one risk factor between 24 and 28 weeks.

- **Two-step strategy**
  - **Glucose challenge test:** For that administer 50 g oral glucose at any time of the day without any regard to food intake. Withdraw blood after 1 hour of 50 g glucose load.

    If plasma glucose is <140 mg% tolerance to glucose can be considered normal and if ≥140 mg% (7.7 mmol/L), further evaluation by glucose tolerance test is required.
  - **OGTT in pregnancy**
    - Carry out in the morning after an 8–14 hours fast
    - Withdraw fasting blood sample
    - Administer 100 g glucose orally in 300 mL of water
    - Withdraw blood at 1-hour interval for 3 hours after glucose load and check the concentration of glucose in the plasma
    - A least 2 values should go beyond the limits shown below `*

Fasting	95 mg%
1 hour	180 mg%
2 hours	155 mg%
3 hours	140 mg%

  - If the results are within normal limits but clinically suggestive of GDM, repeat the test in the third trimester also.
- **One-step strategy**

  Perform, 75 g OGTT (performed in the morning after an overnight fast of 8 hours) with plasma glucose measurements at fasting, 1 hour and 2 hours, at 24–28 weeks of gestation in women not previously diagnosed with overt diabetes. The diagnosis of GDM is made when any of the plasma glucose values are equal to or more than the following values:

Fasting	92 mg%
1 hour	180 mg%
2 hours	153 mg%

Q. 14

Krishnan, 75-year-old senior citizen, a retired bank officer came to Medicine OPD with the following results after a routine check-up.

Urine

Dipstick test for glucose—Positive.

Blood

Random plasma glucose—138 mg% (7.7 mmol/L).

With these results, he was advised to undergo OGTT. The report is given below:

Plasma glucose		Urine—Dipstick test for glucose
Fasting	98 mg% (5.66 mmol/L)	Negative
½ hour	178 mg% (9.9 mmol/L)	Positive
1 hour	168 mg% (9.3 mmol/L)	Positive
1 ½ hours	158 mg% (8.8 mmol/L)	Positive
2 hours	108 mg% (6.0 mmol/L)	Negative

Go through the laboratory results and give your interpretation and the advice.

Ans. 14

Mr Krishnan is suffering from **renal glucosuria**.

The first report showed random plasma glucose 138 mg% which is within normal limits.

But at this plasma glucose value, urine showed positive response to glucose. It can

---

*[Carpenter MW, Coustan DR. Criteria for screening tests for gestational diabetes. Am J Obstet Gynecol. 1982;144:768-73].

happen when the renal threshold for urine diminishes with age. Normally, the renal threshold for urine is 180 mg%. When the renal threshold for glucose is lowered, the glucose starts appearing in urine at lower plasma glucose levels as in this case.

This assumption has to be **confirmed** by OGTT. OGTT showed passing of glucose in urine even when the plasma glucose level is 158 mg% (at 1 ½ hour) which is much lower than the normal renal threshold for glucose.

Q. 15

Meenakshi, 72 years who has been on oral antidiabetic drugs for more than 20 years attended the casualty unit of Medical College with complaints of fatigue, sweating, palpitation and tremors. Upon examining the patient quickly, a sample of blood was withdrawn for plasma glucose estimation and the patient was given 2 doses of 100 mL of 25% dextrose at 15 minutes interval. The patient showed dramatic improvement. The laboratory report is shown below. Give your diagnosis and explanation of the clinical situation.

Plasma glucose: 42 mg% (2.3 mmol/L).

Ans. 15

The patient was suffering from hypoglycemia because of the following reasons:
- She had been on antidiabetic drugs. Her advanced age and longstanding diabetes might have caused impairment of renal function to cause diminished clearance of the antidiabetic drugs. It has led to the over action of glucose lowering effect.
- Symptoms were neuroglycopenic in nature resulting from glucose deprivation of central nervous system.
- Plasma glucose value is suggestive of hypoglycemia.
  The lower limit of fasting plasma glucose is normally around 70 mg% (3.9 mmol/L). Glucose levels between 45 and 50 mg% and with symptoms of hypoglycemia (Whipple's triad) which are relieved promptly after the glucose intake can be documented as hypoglycemia.

Symptoms of hypoglycemia are:
- **Autonomic (adrenergic):** Palpitation, tremor and anxiety
- **Autonomic (cholinergic):** Sweating, hunger and paresthesia
- **Neuroglycopenic symptoms:** Behavioral changes, confusion, fatigue, inability to concentrate, incoordination, irritability, anger and speech difficulty seizure, loss of consciousness.

Whipple's triad constitute the following:
- Symptoms suggestive of hypoglycemia
- Low plasma glucose concentration
- Relief of symptoms upon attaining normal blood glucose level.

Demonstration of a normal plasma glucose concentration in the presence of the above symptoms excludes the possibility of a hypoglycemic insult and search for other causes like anxiety, hyperthyroidism and pheochromocytoma.

Q. 16

A person lying on the street in a comatosed state was brought to the casualty by some people unrelated to the patient. Papers from his pocket showed that he was a diabetic on oral antidiabetic treatment

On examination: Comatosed, intense smell of alcohol.

Blood is taken for preliminary investigations and for plasma glucose estimation.
Plasma glucose—45 mg% (2.5 mmol/L).

What is your diagnosis? Is alcohol consumption related to his illness? Explain.

Ans. 16

Diagnosis: Hypoglycemic coma (alcohol induced) (Fig. 23.1).

Excessive alcohol consumption by a person who had skipped meals can lead to hypoglycemia. During oxidation of alcohol,

Biochemical mechanism of alcohol-induced hypoglycemia

Figure 23.1 Biochemical basis of alcohol-induced hypoglycemia

NAD⁺ is reduced to NADH⁺ + H⁺ causing decreased NAD⁺/NADH+ H⁺ ratio which in turn inhibit the formation of oxaloacetate, thereby diminishing the concentration of OA available for gluconeogenesis. Low rate of gluconeogenesis and reduced glucose intake cause lowering of blood glucose leading to hypoglycemia.

Q. 17

Oral glucose tolerance test (OGTT) performed in different clinical situations are given below as separate line graphs. Identify the clinical situation attributable to each graph line.

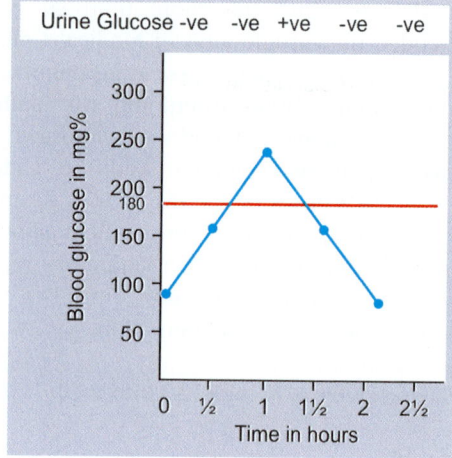

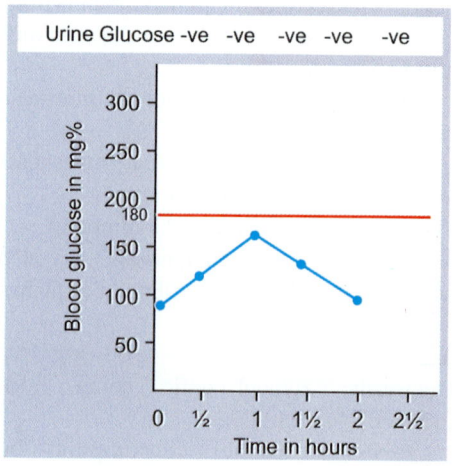

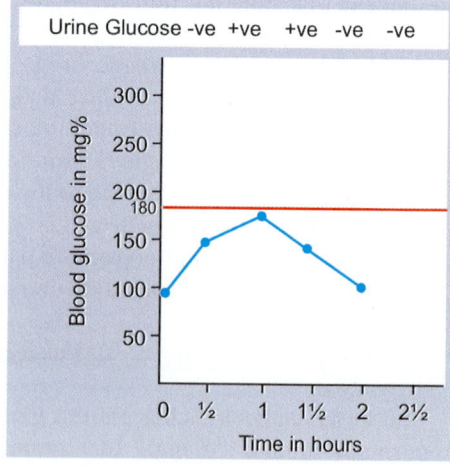

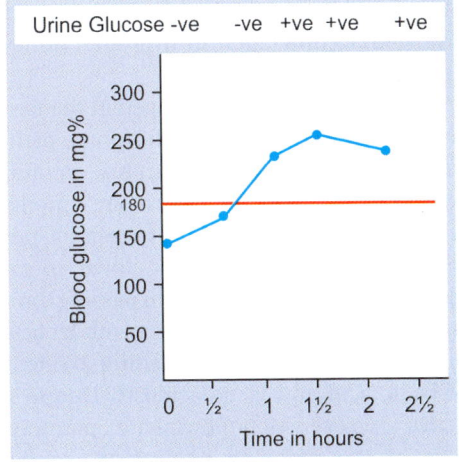

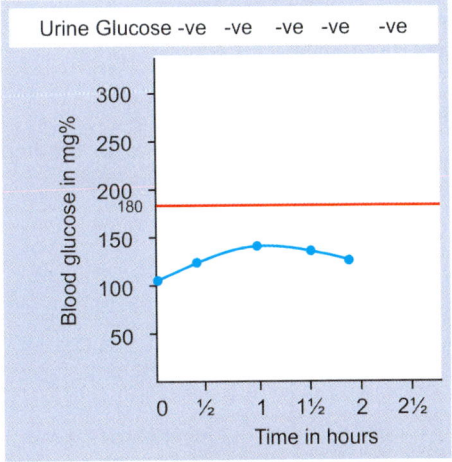

Ans. 17
- **Normal OGTT:** Glucose level within normal limits.
- **Lag curve (Alimentary glucosuria):** Exaggerated rise in blood glucose following an oral glucose load and rapid fall in its level in the blood to touch the normal level at 2 hours. Transient glucosuria occurs due to the peak level crossing the renal threshold for glucose.
- **Renal glucosuria:** It is seen due to lowered renal threshold for glucose, caused by reduced renal tubular reabsorption of glucose.
- **Diabetes mellitus:** FPG level is more than the normal FPG (Normal FPG: <100 mg%) and 2-hour post-load plasma glucose is >200 mg% (Normal 2-hour post-load plasma glucose: <140 mg%).
- **Flat curve (increased glucose tolerance):** Very little rise in blood glucose levels after glucose load. It is seen in patients with hypoactivity of some endocrine glands, e.g. hypopituitarism, Addison's disease and in malabsorption.

INBORN ERRORS OF METABOLISM

Q. 18
A 3-month-old home delivered male baby, brought to the pediatrics OPD On examination Appearance: doll like face with fat cheeks, thin extremities, protuberant abdomen ;

System examination: massive hepatomegaly, enlarged kidneys.

Spleen and heart are not enlarged.

Histopathological examination of biopsy of the liver: distended hepatocytes with glycogen and lipid vacuoles.

Laboratory report		Reference limits for 3-month baby
Blood		
FPG (Fasting plasma glucose)	35 mg%	(60–100 mg%)
ALP	40 IU/L	(40–125 IU/L)
ALT	140 IU/L	(40–125 IU/L)
AST	26 IU/L	(8–20 IU/L)
Uric acid	10 mg %	(2–5.5 mg%)
TG	250 mg%	(30–100 mg%)
Cholesterol	300 mg%	(114–203 mg%)
Lactate	90 mg%	(4.5–20 mg%)

Urine: Dipstick test for glucose—Negative.

Based on clinical and laboratory data what is your diagnosis and explain. Suggest what further tests required for this case?

Ans. 18

Diagnosis: von Gierke's disease (Fig. 23.2).

It is an inherited disorder of carbohydrate metabolism in which **glucose 6 phosphatase enzyme is deficient** in liver, kidney and intestine. This leads to **inadequate conversion of glucose 6 phosphate (G6P) to glucose**. Normally during short fasting, liver glycogen break down to release glucose 6 phosphate which is acted upon by glucose 6 phosphatase to release glucose into the blood to maintain blood glucose levels within normal limits. But in von Gierke's disease due to the deficiency of glucose 6 phosphatase, glucose is trapped in the tissues in the form of glucose 6 phosphate and hence glucose cannot be released into the blood to prevent its level going low. This leads to hypoglycemia after a short fast. Glycogen phosphorylase is inhibited by G6P to produce glycogen accumulation in liver and kidney → enlargement of these organs.

Glucose 6 phosphate trapped in the tissues, pass through Hexose Monophosphate pathway increasingly, so that the surplus amount of NADPH+ H$^+$ and pentoses are produced which favor the excessive synthesis of purine nucleotides and their break down to cause elevated uric acid. The uric acid passing through the glomerular filtrate compete with lactic acid and minimize its excretion leading to retention of lactic acid and lactic acidosis. Due to very active Hexose Monophosphate pathway, a part of the reducing equivalents produced in excess amounts are diverted to reductive synthesis (fatty acid and cholesterol synthesis → hyperlipidemia.

Definitive diagnosis:
- **Liver biopsy:** Enzyme assay in the hepatic tissue; Low glucose 6 phosphatase.

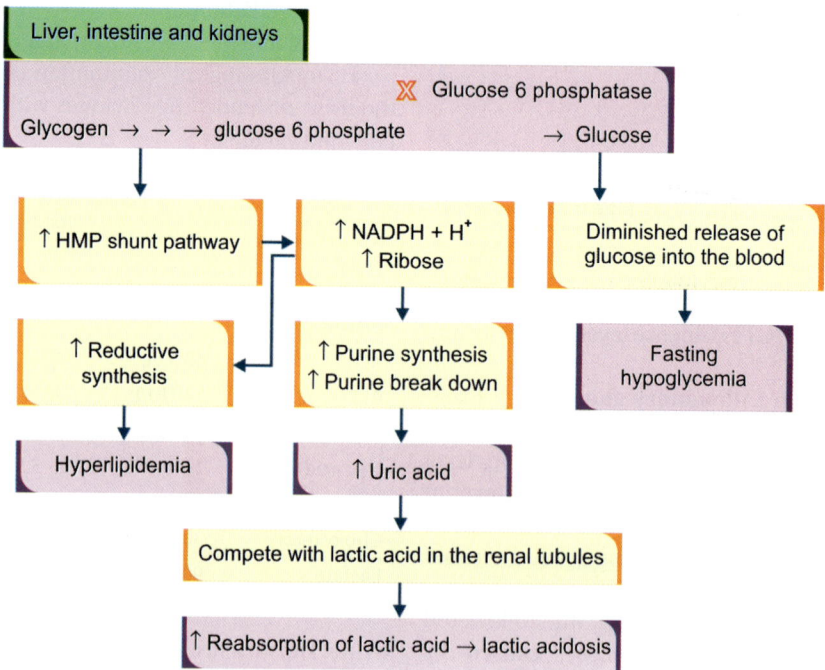

Figure 23.2 Biochemical derangements in von Gierke's disease

HPE: Distended hepatocytes with glycogen and lipid vacuoles.
- **Noninvasive method:** Gene-based mutation analysis.

Q. 19
A male baby aged 6/12 referred from peripheral hospital, presenting with vomiting, jaundice and failure to thrive admitted to pediatric ward of Medical College. Examination revealed cataract in both eyes and hepatomegaly.

Laboratory report

Blood	
Plasma galactose	↑
Plasma glucose	Normal
Urine	
Galactose	Positive

Based on clinical and laboratory data what is your diagnosis?

Ans. 19
Galactosemia: Severe form of galactosemia is due to deficiency of galactose 1 phosphate uridyl transferase leading to accumulation of galactose 1 phosphate in the liver. This will in turn inhibit galactokinase and glycogen phosphorylase. Inhibition of galactokinase will cause accumulation of free galactose and its reduction to dulcitol. Dulcitol due to the osmotic effect causes cataract. The inhibition of glycogen phosphorylase will cause accumulation of glycogen leading to hepatomegaly. The bilirubin uptake and conjugation are affected and will cause unconjugated hyperbilirubinemia and jaundice.

Q. 20
A 30-year-male attended OPD of orthopedics department with case of low back pain, stiffness of knee joint, grayish discoloration of helix of the ear and darkening of urine on exposure to air.

Laboratory report

Urine	
Ferric chloride test	Positive
Benedict's test	Positive

Urine: Developed black color upon keeping in a test tube.

What is your possible diagnosis?

Ans. 20
Diagnosis: Alkaptonuria
It is a autosomal recessive condition due to the deficiency of homogentisate oxidase leading to accumulation homogentisic acid. It undergoes oxidation by polyphenol oxidase to benzoquinone acetate which polymerizes to form black colored alkaptone bodies. These alkaptone bodies get deposited in cartilages and joint surfaces to cause blackish brown discoloration and arthritis.

Q. 21
A 1-year-old female child with case of delayed mile stones, hyperactivity, tremors, convulsions and two hypopigmented patches admitted to pediatric ward for further investigations.

On examination, Delayed mile stones, MR +, mousy odor +.

What is your clinical diagnosis? What all tests are needed to make a diagnosis?

Ans. 21
Diagnosis: Phenylketonuria (PKU)

Blood
Guthrie test
Blood phenylalanine estimation.

Urine
Ferric chloride test

Guthrie's test and ferric chloride test will be positive and blood phenylalanine will be raised to become >20 mg% (Normal blood phenylalanine around 1 mg%).

More about PKU: It is due to deficiency of phenylalanine hydroxylase enzyme as a result of genetic mutation. Due to this defect-phenylalanine cannot be converted to tyrosine and the level of Phe in the blood increases. This disorder results in severe progressive neurological disease due to deficient synthesis of neurotransmitters like serotonin, dopamine and norepinephrine. Phenylalanine passes through aberrant pathways to produce **phenyl ketones** excessively to produce ketosis and mousy odor to the baby. Early recognition of the disease and low phenylalanine diet can minimize the mental retardation.

Q. 22

Mallika, 32 years who is hypopigmented, attended ophthalmology clinic with defective vision and feeling discomfort in the eye, especially during day time.

What is the possible diagnosis and the biochemical abnormality of this disease?

Ans. 22

Diagnosis: Albinism
Normal skin color is due to effect of four biochromes:
1. Melanin (major determinant)
2. Reduced hemoglobin (blue)
3. Oxy-hemoglobin (red)
4. Carotenoids (yellow – exogenous from diet).

Melanin being the major determinant of the skin color, the variations in the amount and distribution of melanin in the skin forms the basis of three common human skin colors—black, brown and white.

Albinism may be due to either absence of tyrosinase (tyrosinase negative albinism) or deficiency of tyrosinase in melanocytes leading to failure of melanin synthesis in the epidermis, hair bulb and eye. It is inherited as autosomal recessive manner. Prevalence 1/20000.

C/F: Skin is white or pink, hair white and pigmentation is lacking in the eye. Poor eye sight, photophobia, nystagmus are common. In the tropics, albinos are prone for photoaging and squamous cell carcinoma at an earlier age.

Diagnosis: Straight forward due to the presence of hypopigmentation.

Histology shows clear cells in the basal layer that fail to stain with Fontana. In tyrosinase, positive albinism dopa reaction is positive (shows deficiency of tyrosinase) and in tyrosinase negative albinism dopa reaction is negative (shows absence of tyrosinase).

Q. 23

A 6-year mentally disabled male child with defective vision was admitted to Ophthalmic ward for further examination.

On examination, mental retardation+ Eyes: myopia, ectopia lentis (dislocation of lens), glaucoma Skeletal system: features of severe osteoporosis + (knock knee, genu valgum, vertebral and foot deformities) Vascular system: Cardiovascular lesions +. Give your provisional diagnosis based on these features. What all supporting biochemical tests you need to confirm the diagnosis?

Ans. 23

Diagnosis: Homocystinuria.

Biochemical Tests
- Silver nitroprusside test on urine sample will give a magenta color if the sample is positive for homocystine
- Thin layer chromatography (TLC)
- High pressure liquid chromatography (HPLC)
- Tandem mass spectrometry
- Molecular genetic testing, e.g. Direct mutation analysis
- Measuring suspected deficient enzyme activities in fibroblasts or in lymphocytes.

More about homocystinuria: Inherited homocystinuria is mostly due to cystathionine β synthase deficiency causing elevated plasma levels of methionine and homocystine and their increased excretion in urine. Homocystine accumulated forms disulfide bonds between two molecules of homocystine to form homocystine which is the predominant form seen in plasma and urine. Accumulation of homocystine causes disruption of collagen and elastin synthesis causing widespread manifestations in the body. Reduced myelination in the brain leads to mental retardation.

Biochemical basis of treatment: Diet low in methionine and rich in cysteine is recommended. Some times pyridoxal phosphate in high doses will relieve symptoms.

Q. 24
A 3-year-old baby referred from peripheral hospital to Pediatric casualty of Medical college with seizures, lethargy, vomiting and a peculiar smell of burnt sugar. Acid base blood analysis revealed acidosis.

What is the possible diagnosis? What all biochemical tests will help you in the diagnosis?

Ans. 24
Diagnosis: **Maple syrup urine disease (MSUD).**

Biochemical Tests
- Dinitrophenylhydrazine test performed on urine specimen will give a yellow white precipitate
- Ferric chloride test—Gray blue color
- Thin layer chromatography (TLC)
- High pressure liquid chromatography (HPLC)
- Tandem mass spectrometry
- Molecular genetic testing, e.g. Direct mutation analysis
- Antenatal diagnosis: Assay of branched chain ketoacid decarboxylase in cultured cells from amniotic fluid.

More about MSUD: It is due to defective **branched chain ketoacid decarboxylase enzyme** deficiency. It is inherited as an autosomal recessive manner. Incidence of this disorder is about 1/200,000. Reduced oxidative decarboxylation of branched chain amino acids (leucine, valine and isoleucine) due to deficiency of branched chain ketoacid decarboxylase enzyme → accumulation of branched chain amino acids and their α-keto or α-hydroxy acid derivatives causing metabolic acidosis.

In the classical type of MSUD, affected infant appears to be normal at birth. Then develop frequent vomiting and failure to thrive. Frequent acute ketoacidotic episodes will occur. Severe neurological features like seizures, coma, respiratory failure can occur and eventually death. Survivors will develop mental retardation.

Q. 25
A Filipino tourist aged 40 years, came to attend a private hospital with complaints of abdominal pain and diarrhea. He said that he had taken milk-based food items on the previous day. He recalled that he had similar episodes in the past upon taking dairy products in mild form. What is your assumption about the disorder, the person suffering from?

Ans. 25
Diagnosis: **Lactose intolerance due to intestinal lactase deficiency**

Biochemical Tests
- **Oral lactose tolerance test**: Patient is asked to come in the morning after a 8–10-hour fast. Fasting blood sample taken for glucose estimation. Patient is given 50 g oral dose of lactose. Blood samples are withdrawn at 5, 10, 15, 30, 45 and 60 minutes after ingestion. An increase in blood glucose concentration of at least 25 mg% over the fasting glucose

concentration will indicate normal intestinal lactase activity. The lactase enzyme acts on lactose to produce glucose and galactose which are absorbed into the blood. Patients with intestinal lactase deficiency will not show any increase in glucose concentration after lactose administration.
- **Direct evidence by histochemical studies of intestinal villous biopsy:** Reveal no lactase activity in cases of lactose intolerance. This is not routinely done.

More about lactose intolerance: Lactase deficiency is the most common disorder of carbohydrate digestion. It may be congenital or acquired. Congenital type manifests during infancy as profuse diarrhea on introduction of milk. Acquired type of **lactose intolerance** is manifested later in life and is common among Ashkenazi Jews, Arabs, Filipinos and Japanese.

Biochemical derangements: Due to lactase deficiency, the lactose ingested remains in the intestinal lumen causing increase in the osmotic pressure of the intestinal contents. Lactose is fermented by intestinal bacteria in the lower intestine to produce gas, lactic acid and fatty acids which also exert osmotic effect causing extrusion of water into the intestinal lumen leading to abdominal distension, flatulence, abdominal pain, diarrhea, dehydration and electrolyte disturbances.

Biochemical principle of treatment: Avoidance of dairy products.

Q. 26
A 1½-year-old child born out of consanguineous marriage presenting with delayed mile stones, convulsions and blindness, admitted in pediatrics ward of a Tertiary care hospital for further evaluation. On examination: Development delay +, prominent rigidity of extensor group of muscles +, increased head size, no hepatosplenomegaly and fundus examination revealed cherry red spot in the macular region of retina. Investigation report showed **lowered activity of hexosaminidase A** in the blood. Comment on.

Ans. 26
Possible diagnosis: Tay-Sachs disease
It is a rare inherited autosomal recessive ganglioside storage disorder. Deficiency of hexosaminidase A → lysosomal deposition of ganglioside in glial and ganglion cells causing cell death.
- Demyelination of white matter and spinal cord leading neurological manifestations.
- Neuron destruction leading to retinal thinning (produces cherry red spot detectable by fundus examination) and blindness.

Tay-Sachs disease is diagnosed by **lowered activity of hexosaminidase A** in the serum, DNA studies and identification of **cherry red spot** in the fundus by ophthalmoscopic examination.

Q. 27
A 40-year-old man referred from periphery to Dermatology OPD as a case of pellagra complaining of headache and unsteadiness of gait for further investigations. On examination: Eczematous skin lesions+, ataxia +. Investigations: Urine amino acid analysis: neutral aminoaciduria. What is your diagnosis?

Ans. 27
Probable diagnosis: Hartnup disease

About Hartnup disease: Incidence: 1/24000; Mode of inheritance: Autosomal recessive.

Defect: Defective neutral amino acid transporter located in the renal tubular and intestinal epithelial cells.

C/F: Pellagra like skin lesions, variable neurological manifestations and passing excessively neutral **amino acids** (alanine, serine, threonine, valine, leucine, isoleucine, glutamine, asparagine) and aromatic (phenylalanine, tyrosine, tryptophan and histidine) amino acids in urine.

Diagnosis: Demonstration of neutral and aromatic amino acids in urine by chromatography.

Q. 28
A boy aged 7 years came for consultation to Pediatric OPD with case of abnormal movements, tendency for self mutilation and joint pains. Investigations showed high uric acid levels in the blood and low activity of hypoxanthine guanine phosphoribosyl transferase (HGPRTase). What is your diagnosis?

Ans. 28
Diagnosis: Lesch-Nyhan syndrome.

About Lesch-Nyhan syndrome: Mode of inheritance: X-linked recessive, affecting males only.

Defect: Deficiency of hypoxanthine guanine phosphoribosyl transferase (HGPRTase) of salvage pathway so that feedback inhibition of de novo synthetic pathway is decreased leading to excessive production and break down of purine nucleotides leading to hyperuricemia.

C/F: Impaired motor function, athetosis and abnormal mental status—violent behavior and self-mutilation.

Q. 29
A man aged 58 years came with history of recurrent attacks of joint pain especially at big toe. On examination, his right metatarsophalangeal joints are swollen and tender.

On investigation, uric acid level was found to be 12 mg%. What is your diagnosis?

Ans. 29
Diagnosis: Gout.

More about Gout: It is a metabolic disease affecting middle aged – elderly men and women due to increased body pool of urate. It is characterized by episodic joint pains due to deposition of monosodium urate crystals in joints and connective tissue (*tophi*) and urate stones in the kidneys (nephrolithiasis). Metatarsophalangeal joint of big toe is first affected.

Investigations:
- Serum uric acid will be elevated
- Aspiration of affected joints or tophaceous deposits—needle-shaped strongly negatively birefringent crystals on polarized light microscopy can be seen (Fig. 23.3).

PORPHYRIAS

Q. 30
Kumar, 40 years was brought by his wife to the outpatient clinic of Medical College with case of recurrent attacks of abdominal pain, headache, palpitation and unpredictable behavior. His wife recalled that his illness used to get aggravated whenever he takes alcohol. The following test

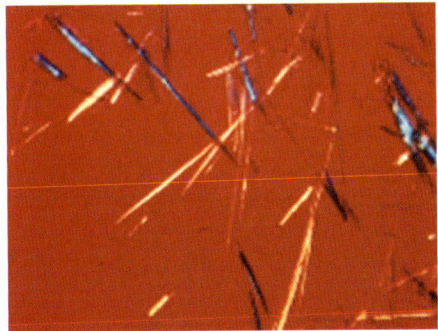

Figure 23.3 Negatively birefringent urate crystals on polarized light microscopy

in the urine was done. Give your most probable diagnosis? Explain the biochemical mechanism of the disease process.

Urine
Watson-Schwartz test for Porphobilinogen—Positive.

Ans. 30
Probable diagnosis: Acute intermittent porphyria due to the deficiency of the enzyme hydroxymethylbilane synthase (Older names – PBG deaminase/uroporphyrinogen 1 synthase).

Porphyrias are rare disorders due to enzyme deficiencies of heme biosynthetic pathway and most often associated with specific mutations which show a autosomal dominant mode of inheritance except in the case of congenital erythropoietic porphyria (CEP). CEP is inherited as autosomal recessive manner.

Mechanism of diseases process: Heme biosynthetic pathway is regulated by product feedback inhibition of the rate limiting enzyme ALA synthase by heme.

When any of the enzymes of the pathway is deficient, heme will not be formed in sufficient amounts to cause feedback inhibition of the key enzyme. Hence, ALA synthase (key enzyme of heme synthetic pathway) activity is uninhibited → production of compounds proximal to the deficient enzyme in excessive amounts → different clinical features in different types of porphyrias.

In this case of **Acute intermittent porphyria**, urine was positive for PBG. In the heme biosynthetic pathway, the PBG is the substrate of the enzyme hydroxymethylbilane synthase (PBG deaminase). From this we could infer that PBG is accumulated in the blood and excreted in urine due to the deficiency of hydroxymethylbilane synthase. This enzyme deficiency is detectable in erythrocytes.

Common precipitating factors include alcohol, exogenously administered steroid drugs and endogenously produced steroids, fasting.

Biochemical principle of treatment is to inhibit the key enzyme, by giving heme.

Q. 31
Revi, 43 years who used to attend the Medical College Hospital from childhood onwards with a progressive disease came to attend medicine OPD due to exacerbation of his illness.
On examination:
  Disfigurement of fingers, nose +
  Reddish brown teeth which fluoresce on exposure to ultraviolet light +
  Skin lesions mostly on exposed parts of the body + Splenomegaly +

Investigations:
Blood: ↑ Uroporphyrin I and ↑ coproporphyrin I

Uroporphyrinogen III cosynthase activity: ↓

Urine: ↑ Uroporphyrin I and ↑ coproporphyrin I
Comment on the type of illness the person suffering from?

Ans. 31
Probable diagnosis: Erythropoietic porphyria
It is due to deficiency of ' uroporphyrinogen III synthase activity. This is inherited as autosomal recessive manner and is the most severe type of porphyria.

The disease can be detected in utero by analyzing the amniotic fluid Porphyrins (type I) and Uroporphyrinogen III cosynthase activity.

Q. 32
Mother noticed pale lips and fatigue, of her 4-year-old baby girl and brought her to a pediatrician. While examining the baby the doctor noticed a toy in the baby's hand which was notorious for high lead content. The mother told that the child was given treatment for intestinal worms and iron supplementation before. Despite these measures, the child

remained pale. On examination: Pallor + No other findings.

Investigations
Urine- delta (δ) aminolevulinic acid: +ve
Blood lead (Pb) level – 50 mg/dL (normal – up to 25 mg/dL)
Erythrocyte Zinc protoporphyrin - Raised
What is your diagnosis? Explain.

Ans. 32
Possible diagnosis: Lead poisoning (Plumbism)
Toxic effects of lead are due to its ability to inhibit ALA dehydratase and ferrochelatase and hence the substrates of these enzymes tend to accumulate. They are ALA and protoporphyrin respectively. Protoporphyrin combine with zinc to form zinc protoporphyrin (ZPP).

C/F: Behavioral changes—irritability, hyperactivity, learning disabilities or generalized symptoms like loss of appetite, nausea, muscle weakness, fatigue, pallor, headache, etc.

Biochemical tests for confirming plumbism:
- Urine—δ - amino levulinic acid
- Erythrocyte zinc protoporphyrin
- Blood lead assay.

VITAMINS AND MINERALS

Q. 33
A women aged 47 years suffering from menorrhagia for the past one year came with case of tiredness, headache. On examination: Pallor + and her hemoglobin was 6 g%. What is the problem with the women? Evaluate.

Ans. 33
Probable diagnosis: **Iron deficiency anemia** due to chronic blood loss for the past one year.

Chronic blood loss by hemorrhage (excessive menstruation/peptic ulcer/hemorrhoids) is the most common cause of iron deficiency among adults. Hookworm infestation is another major cause.

As iron deficiency develops, iron is released from storage compound ferritin. This will continue until iron stores are depleted leading to anemia.

C/F: Distorted appetite (pica) with cravings for ice, earth and in extreme cases spooning of finger nails (koilonychia) may occur and also Plummer-Vinson syndrome (partial occlusion of opening of esophagus).

Biochemical assessment of iron deficiency:
- Ferritn assay in blood—will be lowered (ferritin is the storage form of iron)
- TIBC (Total iron binding capacity)—will be raised.
  (TIBC is a measure of iron binding capacity of transferrin, when fully saturated with iron. Normal range: 250–450 mg/dL).

Q. 34
Pokker, a businessman, aged 52 years came with complaints of weariness, joint pains, darkening of skin color, yellowish discoloration of sclera. On examination: Hyperpigmentation on the face, neck, extensor aspects of lower part of the forearms, dorsal aspects of hands, lower legs +, Hepatomegaly +, splenomegaly +, gynecomastia+, testicular atrophy +.

Investigations	Reference range
Fasting Plasma Glucose: 2-hour post prandial	140 mg%
Plasma glucose: TB	240 mg%
	3.0 mg%
Direct bilirubin mg%	1.2
ALP IU/L	134 (40–125)
AST IU/L	25 (8–20)
ALT IU/L	40 (10-35)
Serum iron mg/dL	200 (50–150)
TIBC mg/dL	150 (250–370)
Transferrin saturation %	90 (22–46)
Serum ferritin mg/L	5000 (20–250)

What is your impression?

Ans. 34
Probable diagnosis: Hemochromatosis

Clinical features and investigations are supporting hemochromatosis.

It is an inherited disorder (autosomal recessive) of iron metabolism in which an inappropriate increase in iron absorption results in deposition of excessive amounts of iron in parenchymal cells leading to tissue damage and later organ dysfunction and are manifested as cirrhosis of liver, diabetes mellitus (due to pancreatic damage), arthritis, hypogonadism and cardiomyopathy.

Genetic basis: It is caused by mutant gene—most commonly by *HFE mutant gene.*

C/F: Hepatomegaly in the absence of any significant symptoms and with only minimal derangement of liver functions.

Excessive skin pigmentation (bronzing) is due to increased melanin and iron in the dermis. **Diabetes mellitus** occurs in more than 50% of the patients which is due to direct damage of pancreatic islet cells by iron deposition.

Arthropathy is also common and reason for its association with hemochromatosis is not clearly known.

Cardiomyopathy: Heart is diffusely enlarged and is prone to develop congestive cardiac failure.

Hypogonadism: Common in both sexes and is manifested as sparse body hair, impotence, gynecomastia and testicular atrophy in males and amenorrhea in females.

Biochemical tests for diagnosis: Ferrokinetic studies as shown below are useful:

- **Plasma iron**—when it is between 180 and 300 mg/dL the disease will be symptomatic
- **TIBC** is decreased in hemochromatosis often approaches zero (Normal range—250–370)
- **Percent transferrin saturation** (Serum iron/TIBC × 100): >50% indicate hemochromatosis
- **Serum ferritin:** >1000 mg/L indicate hemochromatosis.

In this case all the tests done are conforming to hemochromatosis.

Q. 35
A 28-year-old man admitted with cirrhosis, neurological features suggestive of damage to basal ganglia, behavioral changes, Kayser-Fleischer ring in the cornea and sunflower cataracts for further investigation?

Based on these data what is your provisional diagnosis? What biochemical investigations are needed in this case?

Ans. 35
Provisional diagnosis: Wilson disease

Biochemical investigations required:
- Serum ceruloplasmin—Low
- Urine copper—High
- Gold standard for diagnosis—hepatic tissue copper on liver biopsy.

More about Wilson disease: It is an autosomal recessive disorder and is caused by mutations in the *ATP7B* gene which encodes a membrane bound copper—transporting ATP ase. In general its incidence is—1/35000.

Deficiency of ***ATP7B* transporter protein** impairs biliary copper excretion leading to retention of copper in the liver. Accumulated copper exert oxidant damage. Deficient copper incorporation to apoceruloplasmin will cause its premature breakdown leading to low serum ceruloplasmin and low serum copper. As a result, free copper level increases as there is no sufficient copper binding protein ceruloplasmin in the serum to bind with. This will lead to deposition of free copper in various tissues like brain (basal ganglia lesions + psychiatric manifestations, cornea (Keyser-Fleischer ring) and liver cirrhosis leading to hepatic failure.

Q. 36
A few case presentations are given below. Give the possible diagnosis:
- A lady hailing from Tamil Nadu engaged in garbage disposal presenting with edema of legs and face and breathlessness. Upon taking dietetic history, she said that she prefers polished rice only. Biochemical Test done: Erythrocyte transketolase activity—Low.
- A man coming from Rajasthan engaged in construction work came to attend Medicine OPD of Medical College with diarrhea and he also complained about the forgetfulness and unstable mood he had noticed recently. About the diet: He was taking mainly sorghum based diet. On examination: erythematous lesion on the face and feet +, Hyperpigmentation around the neck+, Ataxia +.
- A woman came with complaints of night blindness and dry eyes to Ophthalmic OPD.

On examination: Bitot's spots (patches with dried soap bubble appearance) on the conjunctiva +, Skin is dry and rough. Investigations: Serum vitamin A level—Low; Retinol binding protein—Low.

Ans. 36
- **Wet beriberi**. Eating polished rice for long periods by nutritionally compromised individuals will → niacin deficiency which is known as beriberi. It is of two types:
  (i) Dry beriberi: Mainly neurological manifestations
  (ii) Wet beriberi: Predominantly of cardiovascular manifestations.
  RDA of Thiamine: 0.5 mg/1000 calories (its coenzyme form (TPP) is a cofactor of many enzymes of energy yielding catabolism of nutrients like carbohydrates and fat).
- **Pellagra:** Deficiency of niacin causes diarrhea, dementia and dermatitis.
  Niacin for the body is obtained from the diet and also from tryptophan. Consumption of sorghum aggravates niacin deficiency. Sorghum is rich in leucine. Leucine inhibits quinolinate phosphoribosyl transferase which is the rate limiting enzyme of nicotinamide nucleotide synthetic pathway from tryptophan. (nucleotide form of niacin e.g.: NAD, NADP)
- **Vitamin A deficiency**
  RDA of vitamin A: Children 400–600 µg/day; Adults—750 µg/day

Q. 37
After going through the following case studies give your diagnosis and explanation:
- A 6-year-old female presenting with bow legs, pigeon chest bossing of frontal bones. Her plasma vitamin D level—Low
- A 2-year-old male child admitted to pediatric OPD, with features of bleeding tendency—echymosis, hemorrhage in the mucous membranes. Investigations showed prolonged prothrombin time and delayed clotting time. The condition improved with vitamin K administration.

Ans. 37
- **Rickets due to vitamin D deficiency**
  The active form of vitamin D, calcitriol promotes absorption of calcium and phosphorus from the intestine. Vitamin D deficiency is characterized by inadequate mineralization of bone causing deformation of weightbearing bones.
  Vitamin D is formed mainly from **7-dehydrocholesterol** on exposure to sunlight. Dietary sources are egg yolk, liver and fish liver oils
  RDA ; 400 IU/day (10 µg/day)
- **Vitamin K deficiency**

Vitamin K is needed for the post-translational carboxylation of glutamic residues of proteins like **clotting factors II, VII, IX and X; protein C; protein S; osteocalcin of bone; matrix Gla protein of vascular smooth muscle** to make them functionally active. Hence, deficiency of vitamin K causes defective clotting of blood.

RDA of vitamin K: 50–100 µg/day.

Dietary sources: Green leafy vegetables, margarine and liver (intestinal bacterial flora will also synthesize vitamin K).

Phytoquinone from plants (Vitamin K_1), Menaquinone (Vitamin K_2) from bacterial source and synthetic forms relate to Menadione (Vitamin K_3).

Q. 38

Give explanation for the following:
- Fluoride has got anticariogenic effect
- Iodine deficiency causes goiter
- Vitamin C deficiency causes scurvy
- Phytanic acid oxidase deficeincy causes Refsum disease.

Ans. 38

- **Cariostatic effects of fluoride may be due to the following effects:**
  - Help in apatite crystal formation
  - Stimulation of enamel surface process
  - Decreased enamel solubility
  - Decreased bacterial enzyme activity.
- **Dietary iodine deficiency** (common in hilly areas like Himachal Pradesh) results in decreased secretion of thyroid hormones which by feedback mechanism via thyrotropin causes hyperplasia glandular tissue in an attempt to maintain adequate hormone secretion.

  Some examples for goitrogens: Cassava root which contains thiocyanate (inhibits iodine trapping enzyme of thyroid gland).
- **Ascorbic acid (vitamin C)** acts as a cofactor of the enzyme protocollagen hydroxylase which catalyze hydroxylation of prolyl and lysyl residues of collagen. Hence, deficiency of vitamin C causes inadequate intercellular connective tissue substance. This is manifested as swollen tender bruised skin and, mucous membranes and joints.
- **Peroxisomal phytanic acid oxidase** concerned with α-oxidation of branched fatty acids is deficient in Refsum disease. In the affected individuals, phytanic acid will be accumulated over time, upon consumption of dairy products, meat and fish (originally derived from chlorophyll). Phytanic acid gets integrated into the myelin causing the disintegration of myelin sheath leading to neuronal damage. It is manifested as demyelinating neuropathy, sensorineural deafness, cerebellar ataxia and retinitis pigmentosa causing night blindness.

TUMOR MARKERS

Q. 39

What are tumor markers? Give 5 clinical applications of tumor markers.

Ans. 39

Tumor markers are substances found in increased amounts in the blood, other body fluids or in the tissues and may suggest the existence of a particular type of cancer.

Clinical applications:
- Screening for cancer in asymptomatic individuals
- Clinical staging of cancer
- For estimating disease progression
- Evaluation of the outcome (success) of treatment
- Detection of recurrence of cancer after treatment.

Q. 40

For the following type of tumor markers, give one example of malignancy detectable by each:
- Carcinoembronic antigen

- Alpha-fetoprotein
- CA 125
- Estrogen and progesterone receptors.

Ans. 40
- Carcinoembrionic antigen—colorectal carcinoma
- Alpha fetoprotein—hepatocellular carcinoma
- CA 125 - Endometrial (uterine) carcinoma
- Estrogen and progesterone receptors—Carcinoma breast.

WATER AND ELECTROLYTES

Q. 41
Anto was traveling by bus from Kottayam to Kasargod. He experienced severe vomiting during the journey. Upon reaching Kasargod he felt intense thirst, weakness and dizziness and complained of reduced urine output. He was brought to nearby hospital and was examined by the duty medical officer. On examination: confused, dry mouth, reduced skin turgor and tachycardia + hemogram showed hemoconcentration and high MCV.

Laboratory data:

Plasma sodium	125 mmol/L
Blood urea	54 mg%
Plasma creatinine	1 mg%
Plasma osmolality	230 mOsm/kg
Urine sodium	Decreased

What kind of water and electrolyte imbalance this man suffering from? Explain.

Ans. 41

Diagnosis: Hyponatremia because plasma sodium is <135 mmol/L

Sodium loss occurs in vomiting. General symptoms of hyponatremia like weakness, apathy, lassitude, headache, giddiness and gastrointestinal symptoms like anorexia, nausea and vomiting may occur. Cardiovascular signs-hypotension, tachycardia.

With marked sodium loss—mental confusion, delirium, delusion, stupor or coma may develop

Laboratory findings:
- Hemoconcentration +: Mean corpuscular volume (MCV)—high
- Serum Na^+—low (in spite of hemoconcentration—this help to differentiate sodium loss from water loss where you will get high hematocrit with high serum Na^+ level)
- Blood urea—high (Hyponatremia → decreased osmotic pressure of extracellular fluid (ECF), water moves into the cells causing reduced plasma volume → diminished renal blood flow leading to high blood urea
- Plasma osmolality: A low serum/plasma osmolality indicates low solute concentration (osmolality) of ECF. (**normal plasma osmolality: 275–290 mOsm/kg of water**).

A low serum osmolality can occur in water excess or sodium loss or a combination of both. Serum osmolality can be measured directly by freezing point osmometer or can be found out indirectly using the calculation.

Calculated serum osmolality (mOsm/kg) = 1.86 Na^+ + Glucose/18 + BUN/2.8 (normally the calculated serum osmolality is 5–8 mOsm/kg less than osmolality measured by osmometer).

Conditions causing hyponatremia:
- **Loss of gastrointestinal secretions:** Vomiting, dysentery, diarrhea
- **Loss through skin:** Excessive sweating, exudative skin lesions, burns
- Sequestration of sodium within the body, e.g. Small bowel obstruction—large volumes of intestinal fluid retained in the lumen of the intestine
- **Sodium loss through kidneys:** e.g. Chronic renal disease, syndrome of inappropriate secretion of antidiuretic hormone (SIADH)

- **Continuous secretion of ADH:** Causes retention of water and increased circulating blood volume which in turn causes decrease in the secretion of aldosterone and increase in the secretion of atrial natriuretic factor. This causes kidneys to excrete large amounts of Na^+ in urine leading to hyponatremia
- **Diabetic ketoacidosis:** Loss of glucose and ketone bodies through urine is accompanied by Na^+ leading to hyponatremia
- **Drugs:** For example: Carbonic anhydrase (in the kidney) inhibitors

$$CO_2 + H_2O \leftrightarrow H_2CO_3 \leftrightarrow H^+ + HCO_3^-$$
(Carbonic anhydrase)

 When carbonic anhydrase is inhibited H^+ ions are retained, instead of Na^+ and K^+.
- Severe hemorrhage lead to hyponatremia.

Q. 42
Devaky 40 years was brought to casualty because of vomiting, diarrhea and muscle cramps.

On examination, she was dehydrated. ECG findings: due to delayed depolarization - inversion of T wave, prominent U wave and ST segment depression are seen.

Plama K^+: 2.8 mmol/L; Plasma Na^+: 130 mmol/L.

What is the predominant electrolyte disturbance, the patient suffering from?

Ans. 42
Diagnosis: Hypokalemia (Because plasma Potassium is <3.5 mmol/L).

Potassium and sodium loss occur in vomiting and diarrhea. Here ECG findings suggest hypokalemia. Reference range of Plasma K^+: 3.5–5 mmol/L.

Q. 43
Go through the laboratory data of a patient admitted with diabetic ketoacidosis, given in the table and interpret it.

Laboratory data

pH	FPG (mg%)	Plasma Na^+ mEq/L	Plasma K^+ mEq/L	Plasma HCO_3^- mEq/L	Plasma Cl^- mEq/L	Urine: Rothera's test
7.3	400	135	5.0	20	92	+ ve

Ans. 43
Diagnosis: High anion gap metabolic acidosis

Metabolic acidosis because of low pH, low plasma bicarbonate (being used up for buffering ketoacids) due to diabetic ketoacidosis (plasma glucose 400 mg%).

High anion gap:
Anion gap = $(Na^+ + K^+) - (HCO_3^- + Cl^-)$
= $(135 + 5) - (20 + 92) = 140 - 112 = 28$ mEq/L

It is high due to accumulation of ketoacids. (Normal anion gap is about 12 mEq/L).

KIDNEY DISEASES

Q. 44
Bushra, 5 years old admitted with puffiness of face, weakness and hypertension admitted for further evaluation. Report of relevant investigations is given below. Comment on.

Urine	
Protein	4 g/24 hours
M/E	Granular and epithelial casts +
Blood	
Total protein	4.5 g% (reference range at this age 6–8%)
Albumin	2.0 (reference range at this age 3.8–5.4%)
Cholesterol	300 mg% (reference range at this age 120–200%)
Blood urea	50 mg%
Serum creatinine	1.5 mg%

Ans. 44
Provisional diagnosis: Nephrotic syndrome

The patient has constellation of clinical and laboratory findings suggestive of nephrotic syndrome.

They are edema, hypertension, hypoalbuminemia, proteinuria >3 g per day and hypercholesterolemia.

Characteristic findings that constitute nephrotic syndrome (nephrosis) are heavy proteinuria (>3.5 g/day), hypoalbuminemia, hypercholesterolemia, minimal hematuria, edema and hypertension. This can be caused by different types of renal conditions with evidence of inflammation or without any evidence of inflammation (minimal change nephritis) or with proliferative changes.

Q. 45
Kumar, 40 years who used to have recurrent attacks of sore throat in the past developed rapid onset of headache, malaise, loin pain, hematuria, proteinura, hypertension, diminished urine output and peripheral edema, admitted in the Medicine ward for evaluation. Previous throat swab culture report showed that the sore throat he had in the past, was due to streptococcal infection.

Go through the investigations and comment on.

Investigations

Urine	
Protein	2 g/24 hours
M/E	RBC +++; Pus cells +++
Blood	
Total protein	5 g%
Albumin	3.5 g%
Cholesterol	150 mg%
Blood urea	70 mg%
Serum creatinine	2.2 mg%
ASO titer	↑ (Evidence of β hemolytic streptococci)
C_3 complement	↓ (Consumed in immune complex formation)

Ans. 45
Probable diagnosis: Glomerulonephritis

Streptococcal infection is notorious for causing immune mediated injury to glomerular filtration membrane → damage to GBM (glomerular basement membrane) affecting glomerular filtration consequently renal function.

The patient has the classical features suggestive of acute glomerulonephritis because of sudden onset, hematuria, pyuria, edema, hypertension and oliguria.

Q. 46
Give the usefulness of the following biochemical parameters in assessing renal function:
- Plasma urea and urea clearance
- Serum creatinine and creatinine clearance
- Serum cystatin.

Ans. 46
- **Plasma urea and urea clearance** is influenced by factors not related to the kidney. Urea is the end product of amino acid catabolism. Its level is influenced by the hydration of the body (dehydration causes spurious elevation); protein rich diet, increased protein catabolism, muscle wasting, bleeding into the gastrointestinal tract also cause high urea level.

  Urea is filtered by glomeruli and the some urea upon reaching the renal tubules diffuse out of the tubules passively to re-enter the plasma. This passive back diffusion causes under estimation of glomerular filtration rate, if it is calculated based on urea concentrations in plasma and urine.

  Urea clearance = $[U \times V]/P$
  (where U = urea concentration in urine in mg% ; V = volume of urine;
  P = urea concentration in plasma in mg%)

  Urea clearance (normal) = 75 mL/min which is lower than the true creatinine clearance.

  ∴ *Plasma urea and urea clearance will not give a real status of the kidney disease.*

- **Serum creatinine and creatinine clearance:** Creatinine formed as a result of muscle contraction hence its level is related to muscle mass of an individual. Due to this fact its level remains constant in an individual. Moreover, creatinine is produced endogenously and released into the body fluids at constant rate. But the meat intake will increase the serum creatinine level up to 10%. It is freely filtered at the glomerulus and a small amount is secreted by the proximal convoluted tubule (PCT) and hence creatinine clearance exceed inulin clearance (by a factor of 1.1). Inulin is freely filterable at the glomerular filtration membrane, neither reabsorbed nor secreted by the tubules. **Inulin clearance is considered to be the gold standard procedure for the measurement of GFR.** The drawback of inulin clearance in clinical practice is that it requires exogenous administration.

  Because of this serum creatinine is the most widely used marker for GFR since its value is closer to the true value. GFR is related directly to the urine creatinine concentration and inversely to serum creatinine since, creatinine clearance = $U_{cr}/P_{Cr} \times 24$ urine volume

- **Cystatin C:** It is a cysteine protease inhibitor produced almost at a constant rate by all nucleated cells. Its level is not affected by diet at all. It is constitutively expressed by the cells. It is freely filtered at the glomerulus and not reabsorbed or secreted by the tubules. Because of these reasons it is more specific and sensitive than creatinine.

Q. 47
What is the difference between urine specific gravity and urine osmolality?

Ans. 47
Specific gravity compares the density of a solution (thus the weight and size) to that of an equal volume of water (exact number of solute particles are not taken for calculation).

Osmolality measures exact number of solute particles in solution and is a weight/weight relationship (independent of the size). Osmolality is measured by change in freezing point caused by the number of solutes present in the solution. Therefore, when there are solutes with large molecular weight in urine (proteins, glucose), specific gravity (Normal range 1.003-1.030) will disproportionately increase, but osmolality (Normal range 500-1200 mosm/L) will not. Specific gravity of 1040 may be obtained in a child with proteinuria (e.g. nephrotic syndrome). But this is not attainable by human kidney. Hence, renal concentrating ability is better checked by measurements of osmotic concentration of urine—that is by osmolality of urine using osmometer based on freezing point depression.

BIOCHEMICAL DIAGNOSIS OF MYOCARDIAL INFARCTION

Q. 48
Karim 60 years old, a shopkeeper who had severe chest pain in the morning, brought to the hospital by the fellow workers from the work place in the evening of same day itself. He was admitted in Medicine ward and a blood sample taken for enzyme analysis and the report is given below.

Investigation report
CKMB – 15 U/L (Reference range: 0–4 U/L)
　　LDH – 180 IU/L (Reference range: 100–190 U/L)
What is your inference?

Ans. 48
Provisional diagnosis: The clinical presentation and enzyme studies indicate myocardial infarction.

Patient was brought to the hospital within one day of the onset of chest pain. Myocardial injury causes release of creatine kinase (CK MB) within 3–6 hours of myocardial infarction (MI) from cardiac muscle cells. The high CKMB activity (peak at 24–36 hours) will persist in circulation for 3 days.

(Cardiac troponins are more specific than CKMB).

Activity of LDH will rise slowly from the basal level, only after 10–12 hours after the onset of MI and show peak activity by 48–72 hours and high activity will continue up to 5–10 days.

In this case since the blood sample is drawn within 12 hours of the onset of chest pain → ↑CK MB.

To understand the time course of cardiac marker enzyme activity, see Figure 23.4 and Table 23.1.

Q. 49
Vinayak, 40 years, a businessman upon reaching the airport, seized with sudden onset of chest pain, profuse sweating and nausea. He was rushed to nearest hospital with all modern facilities. It took approximately 75 minutes to reach the hospital. Which cardiac marker you prefer to do at this point of time? Why?

Ans. 49
Preferred marker: Myoglobin

Serum levels of myoglobin rise above the reference interval by 1 hour after the onset of myocardial infarction with peak between 4 and 12 hours. This will be cleared rapidly by kidney within 12 hours.

Points to Ponder:
- If the person had any muscle injury or even an intramuscular injection, the myoglobin will be raised due to its release from skeletal muscle.
- If myoglobin levels remains within normal range within 2–4 hours of the onset of chest pain, recent myocardial injury can be excluded.

Q. 50
What are isoenzymes? What is the role of isoenzymes in the diagnosis of myocardial infarction?

Ans. 50
Isoenzymes are a group of related enzymes catalyzing the same reaction but having different molecular structures and characterized by

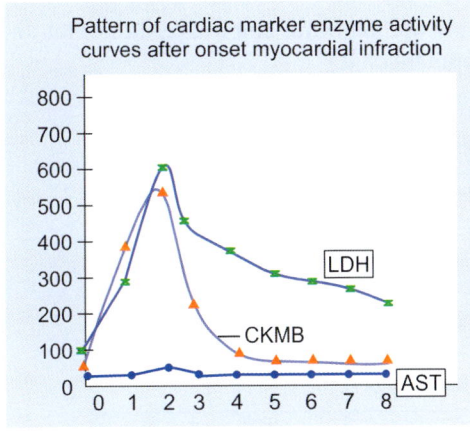

Figure 23.4 Characteristics of cardiac marker enzymes

TABLE 23.1 Time course of cardiac marker enzyme activity			
Enzyme	Time elapsed to exceed the upper reference limit	Peak elevation (hours)	Return to baseline level (days)
Creatine kinase (CKMB)	3–6	24–36	3
Aspartate transaminase	4–8	24	3–4
Lactate dehydrogenase	10–12	48–72	5–10

different physical, biochemical and immunological properties.

Creatine kinase: It consists of 2 subunits—M and B and exists as 3 isoenzymes. They are
(i) CK-1 (CK–BB predominantly in brain).
(ii) CK-2 (CK MB – mainly in myocardium and a small percentage in skeletal muscle).
(iii) CK-3 (CK-MM - mainly in skeletal muscle).

CK-2 (CK-MB) assay is more specific of myocardial injury. Its level start to rise by 3-6 hours, peak at 24-36 hours and come down to basal level by 3 days.

Lactate dehydrogenase catalyzes the reversible reaction between pyruvate and lactate. There are 5 isoenzymes for it (composed of two (M and H) subunits in different proportions):

LDH-1 (H_4 - predominantly in heart)
LDH-2 (H_3 M - mainly in kidney, brain, RBC)
LDH-3 (H_2 M_2 - mainly in spleen, lungs, kidney)
LDH-4 (H M_3 - mainly in spleen, lungs, kidney)
LDH-5 (M_4 - predominantly in skeletal muscle, skin).

Normally serum LDH - 2 is present in greatest amounts.

There is a marked increase in the proportion of LDH-1 level in serum after myocardial infarction. The increase of LDH-1 over LDH-2 in serum after myocardial infarction is called **flipped pattern**. This finding is useful in diagnosing MI.

Q. 51
Abdulla aged 36 involved in an accident, sustained severe crush injury and was on treatment in a hospital. He developed central chest pain radiating to left arm and neck while in the hospital. ECG findings are inconclusive. What type of cardiac marker will be useful in this case?

Ans. 51
Cardiac Troponins (cTn T or cTn I)

CK-2 activity of the heart is 10–20% of total CK activity and that of skeletal muscle is 2%. After skeletal muscle injury, skeletal muscle fraction of CK-2 will increase considerably. In this situation, CK-2 values will not be reliable to diagnose MI. Cardiac troponins is the best available marker in this case.

Cardiac troponin T (cTn T) or Cardiac troponin I (cTn I) will serve this purpose.

cTn T and cTn I increase above the reference limit at 4–8 hours. cTn T and cTn I can remain elevated for 5 days and 10 days respectively.

ENZYMOLOGY

Q. 52
Study the following Lineweaver-Burk Plot and comment on it.

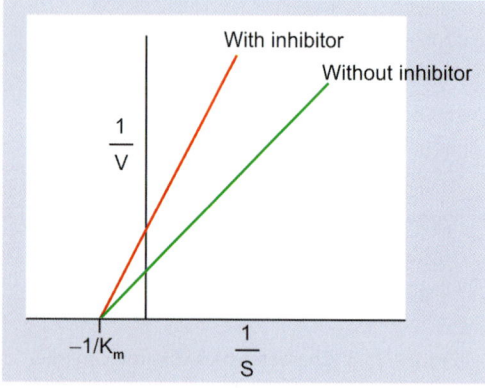

Ans. 52
Noncompetitive inhibition: Inhibitor is not a structural analog of the substrate. It binds to the enzyme at a site other than the active site. The binding causes a conformational change in the structure of the enzyme which in turn alters the active site, so that it cannot bind with

its natural substrate. Noncompetitive inhibition can be reversible or irreversible depending on the type of bond formed. If the inhibitor binds to the enzyme by weak bonds, the inhibition is reversible. But if it is by covalent bonding the inhibition became irreversible. It causes decrease in V_{max} but produces no change to K_m value. Increasing substrate concentration cannot reverse the inhibitory effect.

Clinical significance:
- Mainly poisons act by noncompetitive inhibition, e.g. iodoacetate, cyanide, heavy metal ions (lead, mercury).
- Therapeutic application: British anti-Lewisite used as an antidote for heavy metal poisoning.
- Laboratory use: Inhibition of enolase enzyme of glycolysis by fluoride, is utilized for collecting blood for glucose estimation.

Q. 53
Study the following Lineweaver-Burk Plot and comment on it.

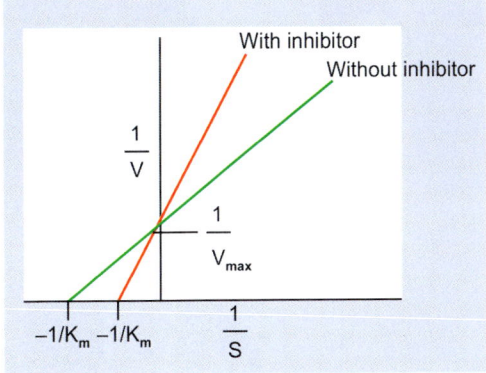

Ans. 53
Competitive inhibition: Inhibitor is a structural analog of the substrate and so the inhibitor competes with substrate for binding to the active site of a specific enzyme. It is possible to reverse the competitive inhibition by increasing the substrate concentration. Here V_{max} of the reaction is not altered but K_m is increased.

Clinical significance:
Therapeutic application: Many widely used drugs act by this principle of competitive inhibition.
Examples:
- Sulfonamides
- Methotrexate.

Q. 54
Study the following Lineweaver-Burk Plot and comment on it.

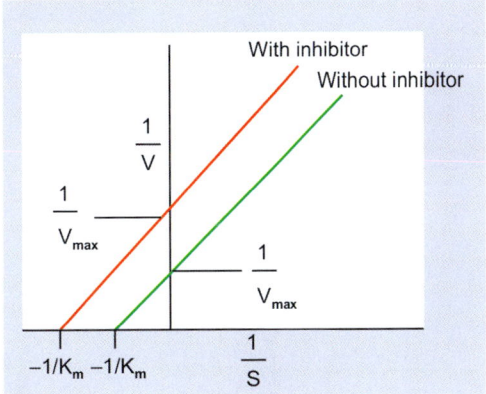

Ans. 54
Uncompetitive inhibition: Inhibitor binds to the ES complex to form an enzyme substrate inhibitor complex that does not generate products. Increase in substrate concentration form more ES complex to which inhibitor binds and wasting the enzyme and substrate without relieving the inhibition.

Application: The principle of uncompetitive inhibition is used in the identification and estimation of placental isoenzyme of alkaline phosphatase (Regan enzyme). In this method, uncompetitive inhibition of placental ALP by phenylalanine is utilized.

SECTION 4

Spotters

CHAPTER 24

Spotters

INSTRUMENTS

Esbach's Albuminometer

Spotter-1

Q. 1
- Identify the instrument.
- What is it used for?
- Briefly note down the procedure adopted.

Ans. 1
- Esbach's Albuminometer.
- Quantitative estimation of protein in urine (approximate).
- Add urine up to 'U' mark and Esbach's reagent up to 'R' mark. Allow to stand for 24 hours. Measure protein precipitate by the scale impregnated on the tube as grams per liter of urine.

Details of Esbach's Albuminometer: It is a glass tube with markings, fitted with a cork. Add urine up to mark 'U' and Esbach's reagent up to 'R'. Allow to stand for 24 hours. Protein picrate would be precipitated and settled at the bottom. The scale impregnated on the tube is calibrated in grams of protein per liter of urine. If the urine specimen is concentrated, dilute it with water to a specific gravity of about 1.008 – 1.010 and follow the same procedure and multiply the reading by dilution factor.

Esbach's reagent: 5 g of picric acid and 10 grams of citric acid dissolved in 500 of distilled water.

electrolyte, subjected to an electric field. If there are differently charged particles they will move in opposite directions. The positively charged will migrate to the cathode and the negatively charged to the anode. The rate of migration of charged particles depends on the charge, mass and shape.

Diagrammatic Representation of Electrophoretic Apparatus

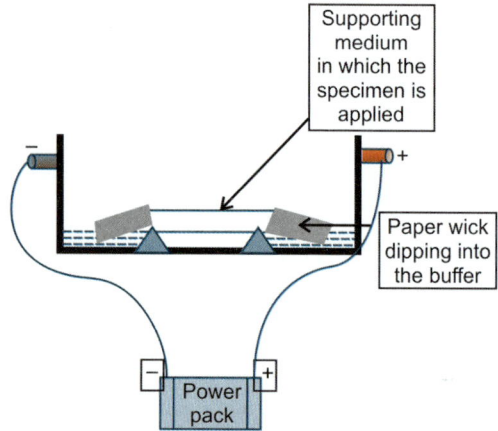

Electrophoretic Apparatus

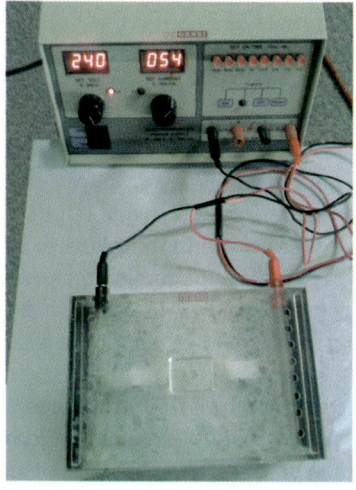

Ryle's Stomach Tube

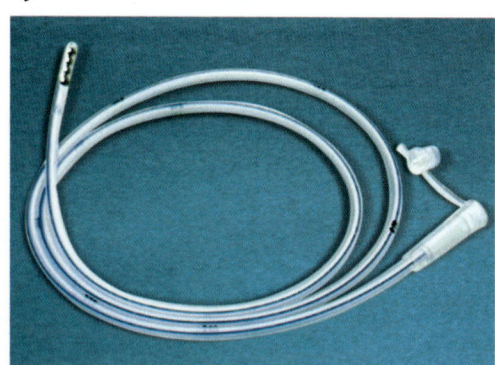

Spotter-2

Q. 2
- Identify the instrument.
- What is it used for?
- Write the principle of the technique used.

Ans. 2
- Electrophoretic apparatus.
- Separation of serum proteins, hemoglobins, RNA, DNA, isoenzymes (e.g. creatine kinase, lactate dehydrogenase).
- **Principle:** It is based on the movement of small charged particles through an

Spotter-3

Q. 3
- Identify the spotter.
- Give two uses of it?
- What is the aim of incorporating lead beads at the closed end?

Ans. 3
- Ryle's tube.
- (1) Aspiration of gastric contents (2) Feeding e.g. In cases of coma, postoperative cases.
- In doubtful cases of misplaced tube, the person can be X-rayed to assess the position of the tip of the tube as the lead beads being radiopaque, cast shadow.

More Details of Ryle's Tube

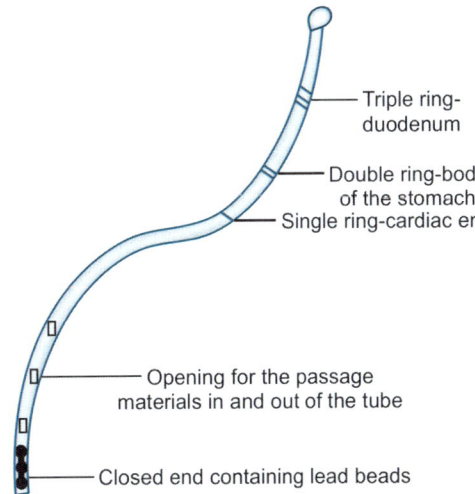

It is a plastic tube of 4 mm external diameter. It has a closed end containing 3 small lead beads and there are holes on the tube, a short distance from the closed end through which the stomach contents can enter the tube during aspiration and leave the tube during feeding. The markings on the tube help to indicate how far the tube has been reached. During introduction of Ryle's tube through mouth, when the single ring marking reaches the lips one can assume that the end has reached cardiac orifice of the stomach. Approximation of the double line with lips indicates that the tip has reached the body of the stomach or almost to the pylorus. Longer tubes contains a triple ring which, when comes in contact with the lips suggest that the end of the tube has entered the duodenum.

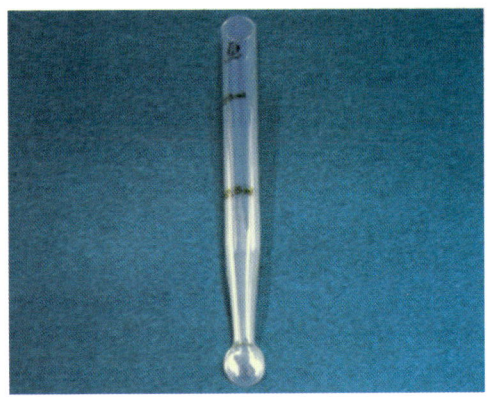

Folin-Wu Sugar Tube

Spotter-4

Q. 4
- Identify the tube.
- What is it used for?
- What is the importance of the constricted portion (neck) of the tube?

Ans. 4
- Folin- Wu sugar tube.
- For blood sugar estimation.
- To minimize reoxidation of cuprous oxide by atmospheric oxygen during the period before adding phosphomolybdic acid (incubation in water bath) Folin and Wu designed the special tube with constricted neck.

More Details of Folin- Wu Sugar Tube
It is a glass tube designed by Folin and Wu to **estimate glucose** by **alkali copper reduction method.** In this method, proteins are precipitated by tungstic acid. Reduction of alkaline copper by sugar in the specimen to form cuprous copper is the principle of the reaction. Then the cuprous copper is oxidized to cupric oxide by molybdic acid which is in turn reduced to molybdenum blue. Final color is measured colorimetrically.

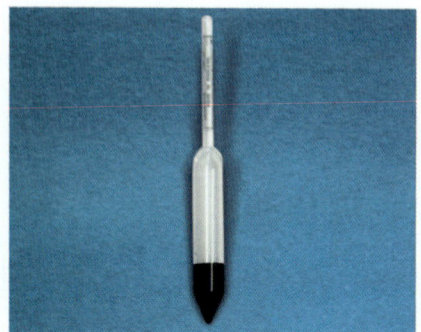

Urinometer

Spotter-5

Q. 5a
- Identify the spotter
- What is it used for?
- Give the temperature corrections to be made on the observed reading.

Ans. 5a

- Urinometer
- To measure specific gravity of urine
- *Temperature corrections:* Add 0.001 for every 3°C rise above 16°C and substract 0.001 for every 3°C dip below 16°C.

Spotter-6

Q. 5b
- Name the instrument used to measure the specific gravity of urine.
- Give the normal range of urine specific gravity and mention two physiological factors affecting it.
- Give one pathological condition each for low and high specific gravity of urine.

Ans. 5b

- Urinometer
- Normal range of urine specific gravity: 1015–1025 (1.015–1.025);
- High water intake causes decrease and excessive perspiration leads to increase in specific gravity.
- High specfic gravity in acute nephritis and low specific gravity in diabetes insipidus.

Spotter- 7

Q. 5c
- Read the specific gravity of the given sample of urine.
- From the value of specific gravity find out the total solids present in the given sample of urine.
- What is meant by fixed specific gravity? Name a condition in which it occurs.

Ans. 5c
- Note the reading on the scale against lowest point of the meniscus: Let it be 1030
- By using Long's coefficient in the urine with specific gravity of 1030
- Solid content in 1 L of urine = 30 × 2.6 × 24 hour urine volume/1000
- Due to loss of concentrating ability of kidney specific gravity is fixed around 1010 and it is called as fixed specific gravity which is observed in chronic kidney disease.

Details of Urinometer

The specific gravity test measures the density of urine relative to the density of water. It varies directly with grams of solutes excreted per liter. It reflects the ability of the kidney to concentrate glomerular filtrate. Specific gravity of urine is determined directly with urinometer. It has a slender neck and a stem with a specific gravity scale usually covers the range from 1.000 to 1.060.

Procedure
- Fill the cylinder about three-fourth full with specimen and place on a level surface.

- Insert the urinometer into the cylinder.
- Read the specific gravity directly from the scale on the stem.
- Take the reading coinciding with the lowest point of the meniscus.
- It is calibrated ordinarily at 16°C (look for the calibration temperature of the urinometer provided).
- Observations made at any other temperature must be subjected to correction to obtain the true specific gravity. For making the correction, add 0.001 for every 3°C rise above 16°C and subtract 0.001 for every 3°C dip below 16°C.

Specific gravity of the urine of normal individuals: 1.015–1.025.

The value fluctuates with water intake and rate of perspiration. It is low with intake of large amount of fluids and it is high with increased perspiration and with low intake of fluids.

Specific gravity of urine in different clinical situations are discussed further.

High Specific Gravity

- Acute nephritis—due to excretion of high amount of proteins in urine (correction for albumin—subtract 0.003 from observed specific gravity reading for each 1 g/dL of albumin in the urine).
- Diabetes mellitus—due to excretion of glucose in urine (correction for glucose—subtract 0.004 from observed specific gravity reading for each 1 g/dL of glucose in the urine).

Low Specific gravity leading to specific gravity less than 1010 (fixed specific gravity)
- Diabetes insipidus—where there is ADH deficiency causing excretion of dilute urine.
- Chronic renal failure—due to inability to concentrate urine.

Calculation of total solids excreted in urine: The total solids excreted in urine may be roughly calculated by **Long's coefficient (2.6)**. The solid content in 1 L of urine is obtained by multiplying the last two figures of specific gravity at 25°C by 2.6, e.g. If the specific gravity of a urine sample is 1.020 and the 24 urine volume is 1200 mL.

Total solids in 24 hr urine = 20 × 2.6 × 1200/1000 = 62.4 g of solids.

A healthy adult excrete around 60 g of solids per liter of urine.

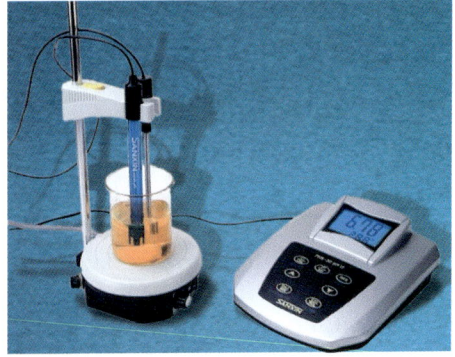

pH Meter

Spotter-8

Q. 6
- Identify the instrument.
- What is it used for?
- Mention the essential parts of the instrument.

Ans. 6

- pH meter
- To measure the pH of fluids
- Internal reference electrode (silver wire dipped in AgCl paste) and unknown electrode sensitive to hydrogen ion concentration.

Details of pH Meter

pH determination is essential in the diagnosis and monitoring of acid-base balance disorders. Use of pH meter allows accurate determination of pH of body fluids and laboratory reagents.

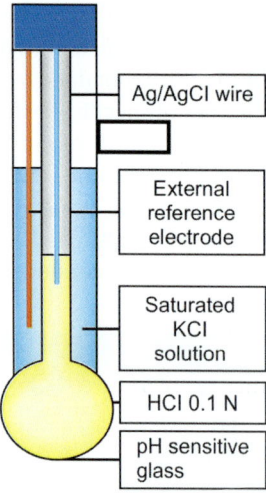

Principle: It is based on the measurement of electromotive force (emf) generated between two electrodes due to difference in [H⁺] concentration. One is reference electrode of known potential and the other is unknown electrode sensitive to hydrogen ion concentration. The electrode potential generated by [H⁺] concentration in an unknown solution is measured against a standard hydrogen electrode potential. A special pH sensitive glass is used in this instrument.

Here the measuring electrode is the glass electrode. Inside the glass is a sliver wire covered with AgCl paste dipped in 0.1 N HCl serving as the internal reference electrode. An electric potential is generated when the thin glass membrane separates unknown solution outside and the known solution inside. The difference in the potential is amplified and converted into direct pH reading on the display unit (digital read out).

Semiautomatic Biochemistry Analyzer

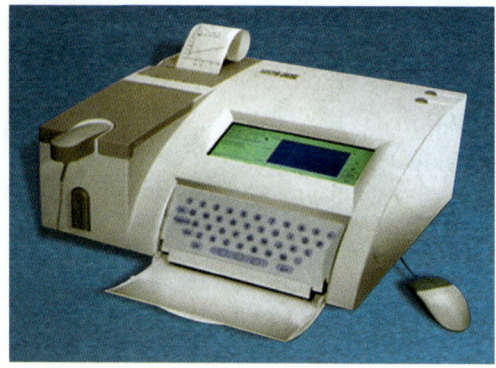

Spotter- 9

Q. 7

- Identify the instrument and give its principle.
- What is it used for?
- Mention the essential parts of the instrument.

Ans. 7

- Semiautomatic clinical chemistry analyzer based on Beer-Lambert law (for details see, Chapter 11-Introduction to Quantitative Analysis).
- For photoelectric colorimetric estimation of compounds in solution.
- Light source, monochromatic filter, photocell, amplifier and galvanometer (for details see Chapter 11-Introduction to Quantitative Analysis).

Dry Chemistry Strip Test for Urine Glucose

Spotter- 10

Q. 8
- Identify the strip.
- What is it used for and give its principle?
- Mention its advantage over Benedict's test.

Ans. 8
- Glucose oxidase based test strips for urine glucose.
- *Use:* To detect and to know approximate amount of glucose in urine.

Principle: The strip is impregnated with peroxidase, glucose oxidase and a chromogenic substrate.

$$Glucose \xrightarrow{Glucose\ oxidase} Gluconic\ acid + H_2O_2$$

$$H_2O_2 \xrightarrow{Peroxidase} H_2O + [O]$$

Colorless Chromogenic substrate $^+[O] \rightarrow$ Colored compound.
The color developed is proportional to the glucose concentration in urine. The color developed is compared with the standard chart provided to know the corresponding glucose concentration.
- *Advantage over Benedict's test:* Glucose oxidase is a specific enzyme which act on β-D glucose only whereas Benedict's test is nonspecific → positive reaction with all the reducing substances present in urine.

REAGENTS

Spotter-11

Q. 9
- Name the ingredients of this reagent and the test done by it.
- Give the role of each ingredient.

Ans. 9
- *Ingredients:* Sodium citrate, sodium carbonate and copper sulfate.
  - Test done: Benedict's test to detect reducing sugars in urine.
- Sodium carbonate—provide alkaline medium; Copper sulfate—yield cupric ions; Sodium citrate—keep the cupric ions in solution.

Spotter-12

Q. 10
- What are the ingredients of this reagent?
- What is the test done by it?

- Mention the use of this test, in the identification of biologically important solutions.

Ans. 10
- *Ingredients:* Copper acetate in acetic acid.
- *Test done:* Barfoed's test.
- *Use:* To differentiate monosaccharides from disaccharides.

Spotter-13

Q. 11
- Name the ingredients of this reagent.
- Give its application in the clinical chemistry.

Ans. 11
- Sodium arsenophosphotungstate in HCl.
- *Application:* It is used for doing Benedict's uric acid test, to detect uric acid in biological fluids.

Spotter-14

Q. 12
- Name the ingredients of this reagent.
- Give its application in the clinical chemistry.
- Give the principle of the test done by it.

Ans. 12
- *Ingredients:* Resorcinol in HCl
- *Application:* For doing Seliwanoff's test which is useful in differentiating ketoses from aldoses.
- *Principle of Seliwanoff's test:* HCl being a weak acid dehydrate ketosugars more readily than aldose sugars to form furfural or furfural derivatives which in turn condense with resorcinol to give a red colored complex.

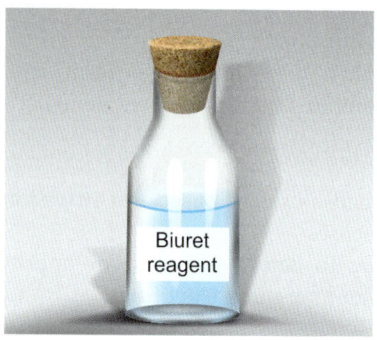

Spotter-15

Q. 13
- What is the composition of this reagent?
- What is its use in the clinical chemistry laboratory?

Ans. 13
- *Composition:* Sodium potassium tartrate, cupric sulfate, potassium iodide and sodium hydroxide in water.
- *Use:* Estimation of serum proteins.

Spotter-16

Q. 14
- What is the composition of this reagent?
- Name two tests done by it.

Ans. 14
- *Composition:* α-naphthol in alcohol.
- *Tests done:*
  a. Molisch test, a general test for carbohydrates
  b. Rapid furfural test to detect ketosugars which is useful to differentiate ketoses from aldoses.

Spotter-17

Q. 15
- Name the ingredients of this reagent.
- Mention its use.

Ans 15
- *Ingredients:* Trichloroacetic acid and ferric chloride in water.

- *Use:* To perform Modified Fouchet's test to detect bilirubin in urine.

Spotter-18

Q. 16
- Name the ingredients of this reagent.
- What is its use?

Ans. 16
- *Ingredients:* Para-dimethylaminobenzaldehyde in hydrochloric acid.
- *Use:* It is used to perform Ehrlich's test to detect urobilinogen in urine.
  Ehrlich's test will be strongly positive in hemolytic jaundice.

INDICATORS

Spotter-19

Q. 17
Mention the pH range and color change and use of this indicator in the clinical chemistry laboratory.

Ans. 17
pH range: 3.8–5.4; **Color change:** Yellow to green.

Use: Isoelectric precipitation of casein at pH 4.6.

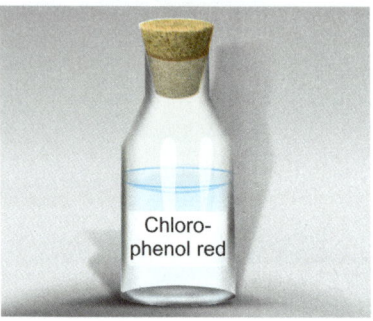

Spotter-20

Q. 18
Mention the pH range, color change and use of this indicator in the clinical chemistry laboratory.

Ans. 18
pH range: 4.8–6.8; **Color change:** Yellow to red.

Use: Heat coagulation test for albumin—Isoelectric precipitation of albumin.

Spotter-21

Q. 19
Mention the pH range, color change and use of this indicator in the clinical chemistry laboratory.

Ans. 19
pH range: 1.2–2.8 (acidic range); **Color change:** Red to yellow. **Use:** Specific test for sucrose.

Spotter-22

Q. 20
Mention the pH range, color change and use of this indicator in the clinical chemistry laboratory.

Ans. 20
pH range: 6.8–8.4; **Color change:** Yellow to red. **Use:** To provide the optimum pH of 6.8 for the urease enzyme used in the specific urease test.

CRYSTALS

Spotter-23

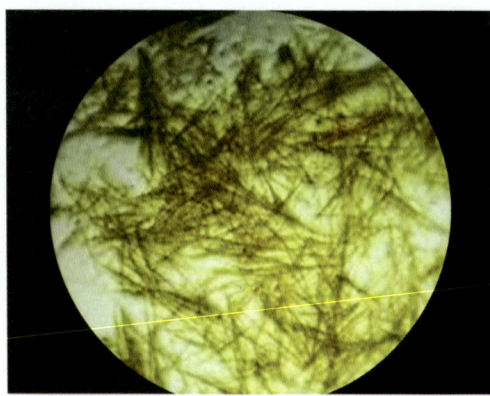

Q. 21
- Identify the crystal.
- Name three biologically important compounds giving this type of crystal.
- Name the test done.

Ans. 21
- Needle-shaped osazone crystal.
- Glucose, fructose and mannose.
- Osazone test.

Spotter-24

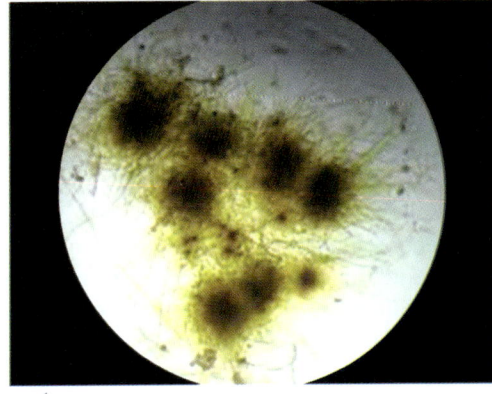

Q. 22
- Identify the crystal and name the compound giving this type of crystal.
- Mention three conditions in which you may get this compound in urine.

Ans. 22
- Puff-shaped lactosazone crystals given by lactose.
- Lactose may be excreted in urine
  - During lactation
  - In third trimester of pregnancy
  - In neonates.

Spotter-25

Q. 23
- Identify and describe the shape of the crystal.
- Name the compound giving this type of crystal.

Ans. 23
- Maltosazone crystal—Individual crystals of maltosazone look like a yellow colored petal, when grouped, looks like a sunflower.
- Maltose.

Spotter-26

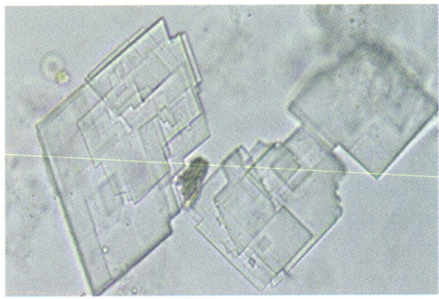

Q. 24
- Identify the crystal.
- Name three biologically important compounds derived from it in the body.

Ans. 24
- Cholesterol crystals—Rhombic crystals notched at one corner.
- Bile acids, vitamin D and steroid hormones.

Spotter-27

Q. 25

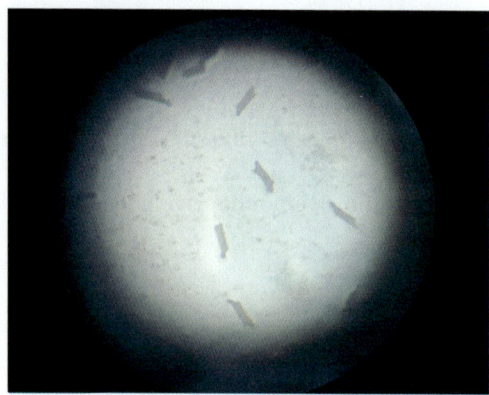

- Identify the crystal.
- Describe the chemical nature of the crystal.
- Mention its application?

Ans. 25
- Hemin crystals (Rhombic shaped).
- Hemin—chloride of hematin (Hematin is ferric iron + protoporphyrin).
- *Application:* Mainly in the forensic medicine to differentiate between blood stain and other stains.

SEPARATION TECHNIQUES

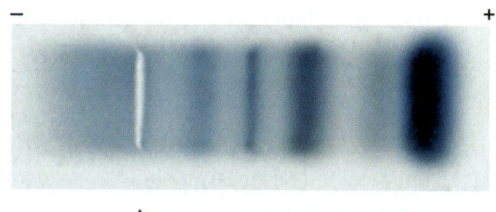

↑ Point of application

Spotter-28

Q. 26
- Identify the strip and the bands.
- Name five biologically important substances separated by this technique.
- Give the principle of the technique used.

Ans. 26
- Electrophoretogram of serum proteins.
  Bands read from the anode: Albumin, α-1 globulins, α-2 globulins, β-1 globulins, β-2 globulins and γ-globulins as shown below:

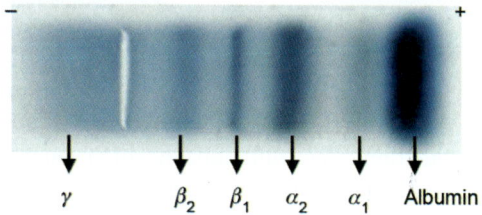

γ β_2 β_1 α_2 α_1 Albumin

- Biologically important substances separated by electrophoresis:
    - Serum proteins
    - Isoenzymes (e.g. Lactate dehydrogenase, creatine kinase)

- DNA
- RNA
- Lipoproteins.
- **Principle:** It is based on the movement of small charged particles through an electrolyte subjected to an electric field. If there are differently charged particles they will move in opposite directions. The positively charged will migrate to the cathode and the negatively charged to the anode. The rate of migration of charged particles depends on the charge, mass and shape.

Spotter-29

Q. 27

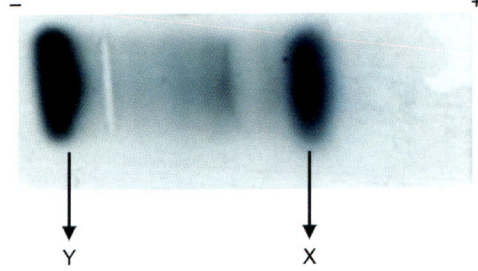

- Identify the strip.
- Name the bands marked as X and Y and comment on.

Ans. 27
- Serum protein electrophoretogram, showing a monoclonal band at the gamma region
- X—Albumin; Y—Monoclonal band (Paraprotein)

Most common cause for producing monoclonal band is multiple myeloma.

More about paraproteinemia (Fig. 24.1): Normally several different clones of plasma cells produce immunoglobulins so that a faint broad band is obtained at the gamma region on electrophoresis. Increase in immunoglobulins can produce two different types of electrophoretic separation patterns. Infections can stimulate a variety of clones of plasma cells which produce different immunoglobulins leading to an appearance of a diffuse band on serum protein electrophoresis which is referred to as **polyclonal band**.

But if a single clone of malignant plasma cells proliferate as in multiple myeloma to produce a single type of Ig (immunoglobulin) or a part of Ig (kappa or lambda light chain), cause an intensely staining band which is more common in the gamma region and is referred to as **"M" (monoclonal band)**.

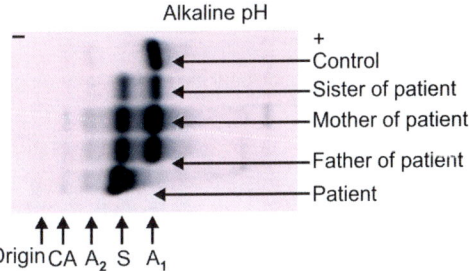

Spotter-30

Q. 28

A male child aged 7 years, exhibited symptoms and signs of hemolytic anemia and the sickling test was found to be positive. Then the whole family screened by Hb electrophoresis against a control. Interpret the Hb electrophoretogram. (CA – carbonic anhydrase; A_2 – hemoglobin A_2; S – hemoglobin S; A_1 – hemoglobin A_1).

Ans. 28
- Patient is suffering from homozygous sickle cell disease, because electrophoretogram indicates the presence of Hb S only and no other bands suggestive of Hb A_1.
- Father, mother and sister are heterozygous for Hb S since both Hb S and Hb A_1 are separated out upon electrophoresis.

Spotter-31

Q. 29

Identify and comment on.

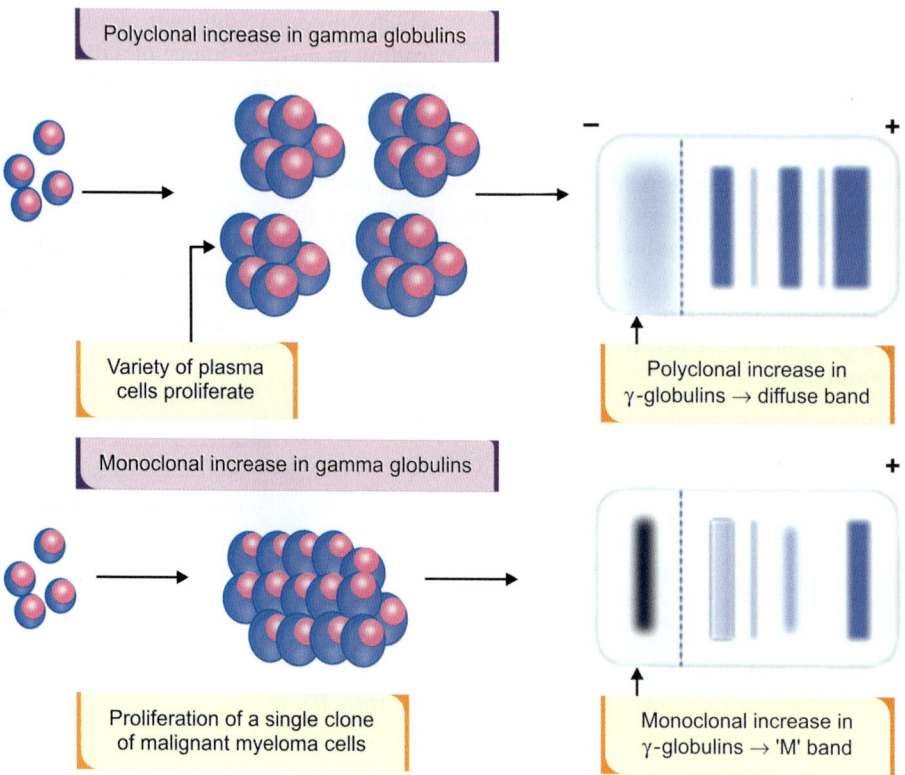

Figure 24.1 Two types of hypergammaglobulinemia—polyclonal and monoclonal

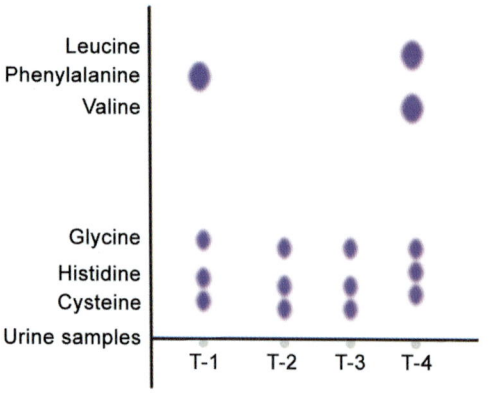

Ans. 29

Urine paper aminoacidogram (Urine amino acid paper chromatogram).

- T-1 – Suggestive of phenylketonuria (due to phenylalanine hydroxylase enzyme deficiency) because a spot suggestive of Phe is seen.
- T-2 – Normal (only the amino acids which are excreted normally are seen).
- T-3 – Normal (only the amino acids which are excreted normally are seen).
- T-4 – Maple syrup urine disease due to branched chain amino acid decarboxylase deficiency since spots indicative of branched chain amino acids (leucine and valine) are visualized on the chromatogram.

More about paper chromatography

Principle: It involves physical method of separation in which the components (solutes) of a mixture are separated by their differential distribution between stationary and mobile phases.

- After chromatographic run, distance traveled by the solvent and the solute along the support medium is measured from the point of application. The ratio of distance traveled by the solute to the distance traveled by the solvent is a constant for a particular solute under particular laboratory conditions. The ratio is called **R_f value (ratio of fronts)** of the amino acid which is specific for the amino acid and hence useful in the identification of amino acids.

$$R_f \text{ value} = \frac{\text{Distance traveled by an amino acid from the point of application}}{\text{Distance traveled by the solvent}}$$

Routine application: Paper chromatography is used routinely in the clinical chemistry laboratory for analyzing urine for aminoaciduria (separation and identification of amino acids, e.g. Phenylketonuria, maple syrup urine disease, homocystinuria) and glycosuria (separation and identification of sugars).

Types of chromatography based on the mechanisms applied: Partition chromatography, Ion exchange chromatography, adsorption chromatography, affinity chromatography.

GRAPHS

Spotter- 32

Q. 30
Identify the graph and comment on.

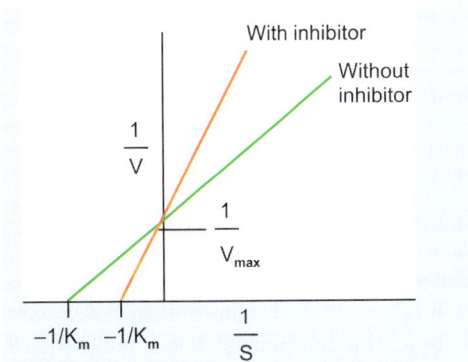

Ans. 30
Lineweaver-Burk plot showing competitive type of enzyme inhibition because here V_{max} is not changed but K_m is increased.

Application in clinical medicine:
- **Drug designing**—examples of drugs used, based on the principle of competitive inhibition are sulfonamides and methotrexate.
- **Treatment of methanol poisoning** by ethanol is based on competitive inhibition.

Alcohol dehydrogenase
Ethanol ⟶ Acetaldehyde
Methanol ⟶ Formaldehyde (highly toxic → optic neuritis leading to blindness)

As shown above the same enzyme alcohol dehydrogenase is concerned with metabolism of ethanol and methanol. When methanol is ingested it will be metabolized to toxic formaldehyde. To counteract this ethanol is given to compete with methanol for alcohol dehydrogenase so as to decrease the conversion of methanol to formaldehyde.

Spotter-33

Q. 31
Identify the graph and comment on.

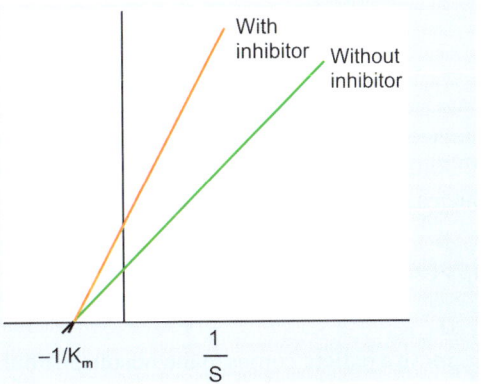

Ans. 31
Lineweaver-Burk plot showing noncompetitive inhibition because here V_{max} is decreased but there is no change in K_m.

Clinical Significance
- Mainly poisons act by noncompetitive inhibition, e.g. iodoacetate, cyanide, heavy metal ions (lead, mercury).
- *Therapeutic application:* British anti-Lewisite used as an antidote for heavy metal poisoning.
- *Laboratory use:* Inhibition of enolase of glycolysis by fluoride is utilized while collecting blood for glucose estimation.

Spotter-34

Q. 32
Identify the graph.

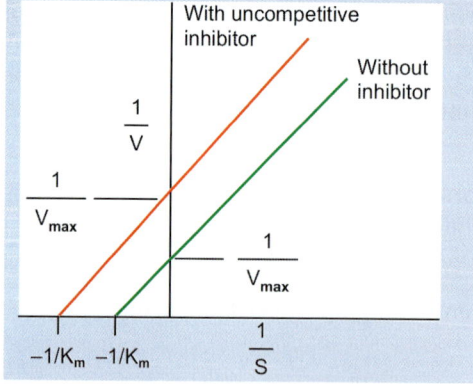

Ans. 32
Lineweaver-Burk plot showing uncompetitive inhibition because here both V_{max} and K_m are decreased.

Spotter-35

Q. 33
Suppose a patient comes to the hospital 6 hours after the onset of chest pain.

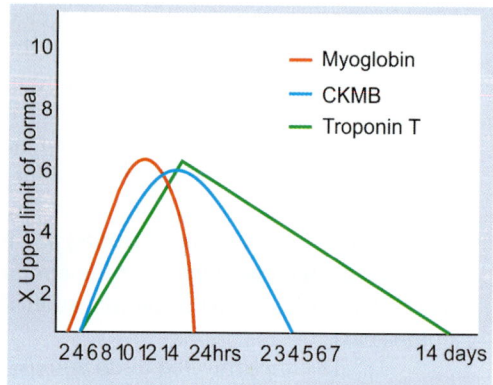

- Which enzyme assay will be useful to diagnose myocardial infarction in the first 6 hours of onset of chest pain?
- Name a nonenzyme cardiac marker which is not influenced by muscle injury.

Ans. 33
- Creatine kinase–MB (CK-MB) the isoenzyme of creatine kinase.
- Nonenzyme cardiac marker: Troponin T (it is not elevated in muscle injury).

Spotter-36

Q. 34

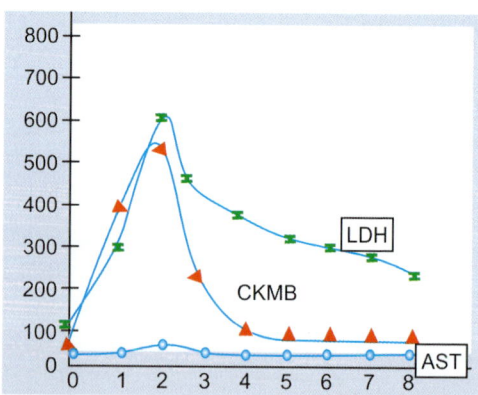

A diabetic patient who had been feeling uneasiness for the past 4 days came for cardiology consultation. The cardiologist suspected ischemic heart disease (IHD) in the patient. (Because

longstanding diabetes will cause neuropathy due to which pain of MI may not be felt).
- Which enzyme will help to diagnose IHD on the 5th day?
- Mention the more specific isoenzyme of it useful in this context.
- What is meant by flipped pattern?

Ans. 34
- Lactate dehydrogenase
- LDH-1
- Normally, serum LDH-2 is the predominant isoenzyme in serum.

There is a marked increase in the proportion of LDH-1 level in the serum after myocardial infarction. The increase of LDH-1 over LDH-2 in the serum, after IHD is called **flipped pattern**. This finding is useful in the diagnosis of IHD.

Spotter-37

Q. 35

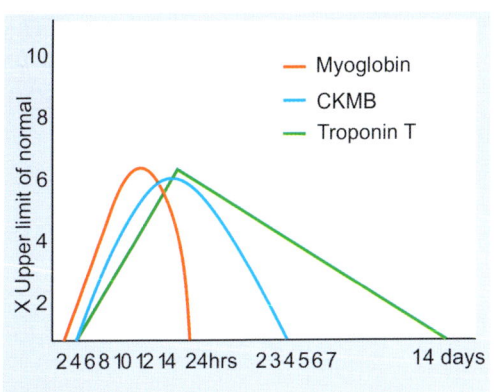

- Is there any marker enzyme, which is useful in the diagnosis of myocardial infarction in the first one hour after the onset of chest pain?
- If no enzyme marker is available, name any other marker useful in the first hour.

Ans. 35
- No enzyme marker is available to serve as a cardiac marker of myocardial infarction in the first hour after the onset of chest pain.
- Myoglobin is useful in the first hour.

Spotter-38

Q. 36
- Identify the graph and interpret.

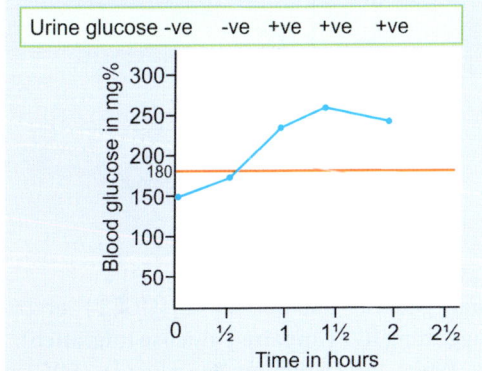

Ans. 36
Oral glucose tolerance test (OGTT) graph showing the curve, indicative of diabetes mellitus.

Fasting plasma glucose (FPG) > 126 mg% (7.0 mmol/L) and 2-hour post-load glucose level > 200 mg % (11.1 mmol/L) is diagnostic of diabetes mellitus.

In this case, FPG is 148 mg% (8.2 mmol/L) and 2-hour post-load glucose level 220 mg% (12.2 mmol/L).

Points to Ponder: To convert glucose value in mg% to mmol/L divide the value in mg% by 18

$$\text{Glucose mmol/L} = \frac{\text{Glucose in mg\%}}{18}$$

Spotter-39

Q. 37
Identify the graph and give your opinion.

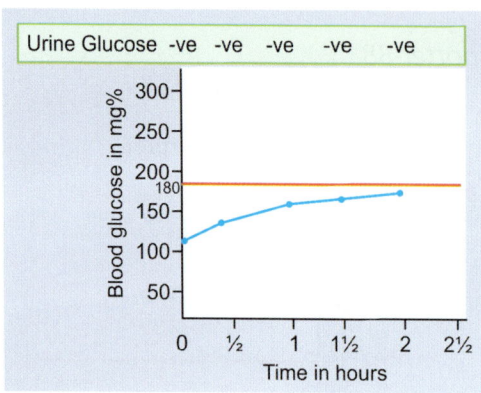

Ans. 37
Oral glucose tolerance test (OGTT) graph, suggesting **IGT (impaired glucose tolerance)**.

When the fasting plasma glucose (FPG) is less than 126 mg% and 2-hour post-load glucose level between 140 and 199 mg%, it is suggestive of impaired glucose tolerance (IGT). In this case, FPG is near 115 mg% and 2-hour post-load glucose level is 170 mg%.

Spotter-40

Q. 38
Identify the graph and interpret.

Ans. 38
Oral glucose tolerance test (OGTT) graph showing.

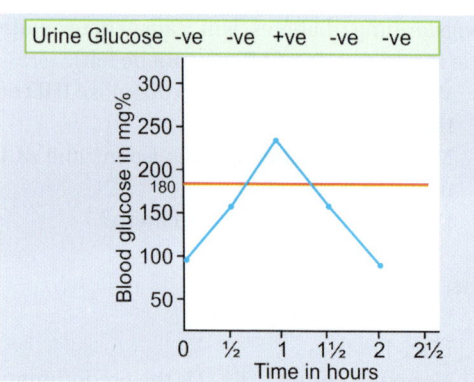

Lag curve suggesting alimentary glucosuria. Exaggerated rise in blood glucose following an oral glucose load and rapid fall in plasma glucose level to touch the fasting level at 2 hours. Transient glucosuria occurs due to the peak glucose level crossing the renal threshold for glucose.

Spotter-41

Q. 39
- Identify the graph and comment on it.
- What are the indications for doing OGTT?
- What is the glucose load to be given for conducting OGTT?

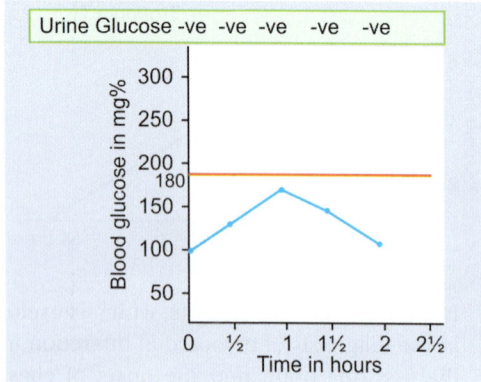

Ans.39
- **OGTT** graph showing **normal glycemic status**
- **Indications** for OGTT
  - For diagnosing gestational diabetes mellitus
  - For diagnosing impaired fasting glucose (IFG)
  - For diagnosing impaired glucose tolerance (IGT)
  - To rule out benign renal glucosuria.
- **Glucose load:**
  - **75 g glucose** (anhydrate form) in adult males and nonpregnant adult females
  - **1.75 g/kg body weight** for children (maximum of 75 g)
  - **100 g** in pregnant women.

  The prescribed amount of glucose is to be taken by dissolving in 300 mL water and should be consumed within 15 minutes.

Spotter-42

Q. 40
Identify the graph and comment on it.

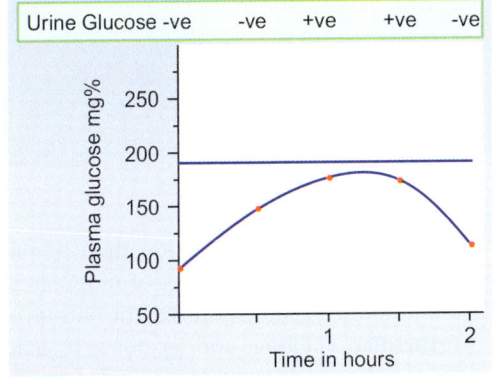

Ans. 40
Oral glucose tolerance test (OGTT) graph suggestive of renal glucosuria because all the plasma glucose values are within normal limits but the test for glucose is positive in two samples indicating renal glucosuria due to lowered renal threshold for glucose which is around 180 mg% in a normal person.

Spotter-43

Q. 41
Identify the graph and comment on it.

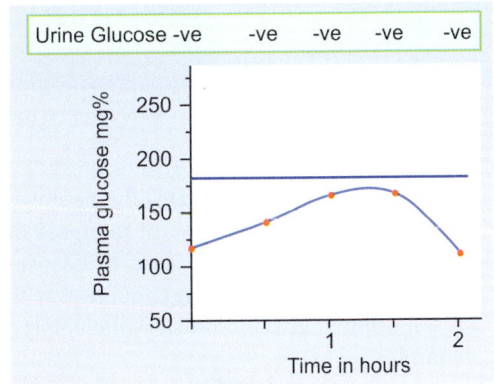

Ans. 41
Oral glucose tolerance test (OGTT) graph showing **impaired fasting glycemia (IFG)**.
- Impaired fasting glycemia is said to occur when the plasma glucose level is above 100 mg%* but below the cut off value used to designate diabetes mellitus 126 mg% (that is between 100 and 126 mg%).
- In this case, fasting glucose is 124 mg% and other values are within normal limits.

TESTS

Spotter-44

Q. 42
- Identify the test.
- Give its principle and its application.

---
* *ADA recommendation in 2003: Lowered the lower limit of Impaired Fasting Glucose from 110 to 100 mg%*

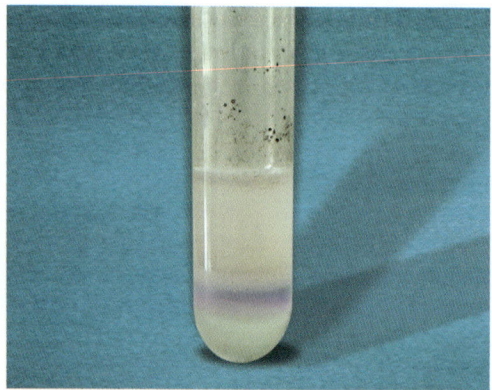

Ans. 43
- Aldehyde test (Identification tips: Violet colored ring and the presence of protein coagulum in the top layer).
- **Principle:** Mercuric sulfate in sulfuric acid act as an oxidizing agent and it oxidizes the indole ring of tryptophan. Then formaldehyde reacts with the oxidized indole ring to form purple colored complex.

Application of the test: Used to detect indole ring containing amino acid, Tryptophan.

Ans. 42
- Molisch test.
- Concentrated acid dehydrates the pentoses to form furfural in the case of pentoses or furfural derivatives in the case of hexoses and heptoses which in turn condenses with α-naphthol to give a reddish violet colored complex.

Application of the test: Used as a general test to detect the presence of carbohydrates.

Spotter-45

Q. 43
- Identify the test.
- Give the principle and its use in the laboratory.

Spotter-46

Q. 44
- Identify the test.
- Give its principle.
- Mention its application in clinical medicine.

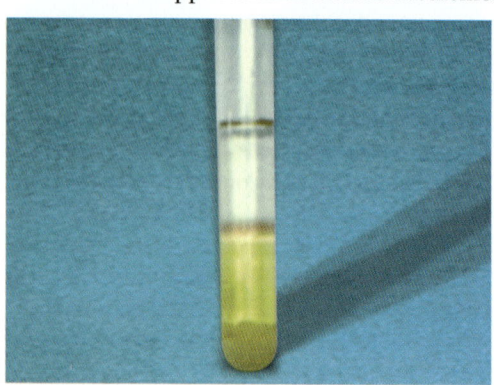

Ans. 44
- **Rothera's test (Identification tips):** Violet colored ring and the presence of undissolved ammonium sulfate crystals at the bottom).
- **Principle:** Acetone and acetoacetic acid react with sodium nitroprusside (nitroferricyanide) in the presence of alkali to produce a purple color.
- **Application of the test:** Used to detect ketone bodies (acetoacetic acid and acetone) in urine, e.g. Diabetic ketoacidosis, starvation ketoacidosis.

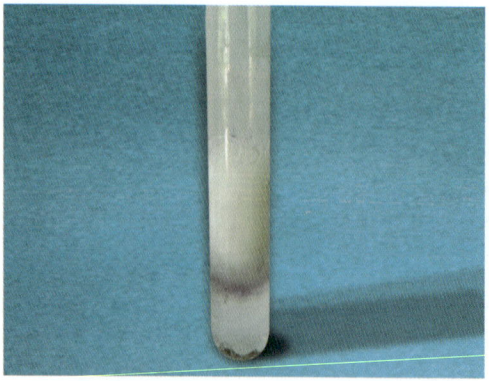

Spotter-47

Q. 45
Identify the test and comment on it.

Ans. 45
Benedict's test. It is a semi-quantitative test. The color of the precipitate indicates the concentration of sugar in solution. Here the color of the precipitate is orange which approximately corresponds to the concentration of 1.5 g%. Carbohydrates with a free aldehyde or keto group reduce various metallic ions. In this test, cupric ions are reduced to cuprous ions by the enediols formed from sugars in the alkaline medium of Benedict's reagent.

Application of the test: Used to differentiate between reducing and nonreducing sugars.

Spotter-48

Q. 46
- Identify the test.
- Give the principle and its use in the laboratory.

Ans. 46
- Barfoed's test.
- **Principle**: It is a reduction test. Reducing property is due to the carbonyl group (aldehyde or keto group) of sugars. Barfoed's reagent is copper acetate in acetic acid. In the acid medium, monosaccharides enolize much more readily than disaccharides and these enediols thus formed reduce cupric ions released by copper acetate to cuprous ions.

Application of the test: Used to differentiate monosaccharides from disaccharides.

Spotter-49

Q. 47
- Identify the test.
- Give the principle and its use in the laboratory.

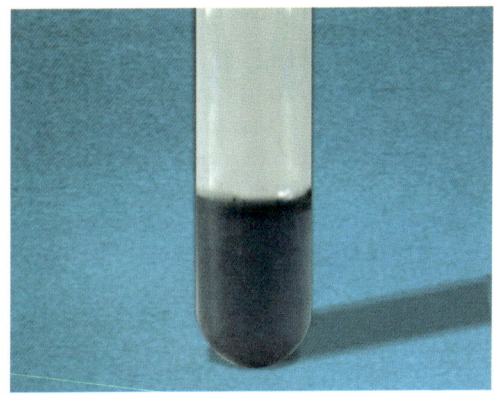

Ans. 47
- Rapid furfural test.
- **Principle**: A dehydration reaction due to the hydroxyl groups of the sugar.
  Concentrated HCl being weaker than concentrated sulfuric acid, dehydrate ketoses (e.g. fructose) more readily than aldoses to form hydroxymethyl furfural which then condenses with α-naphthol to form a violet colored complex.

Use of the test: Used to differentiate ketoses from aldoses.

Spotter-50

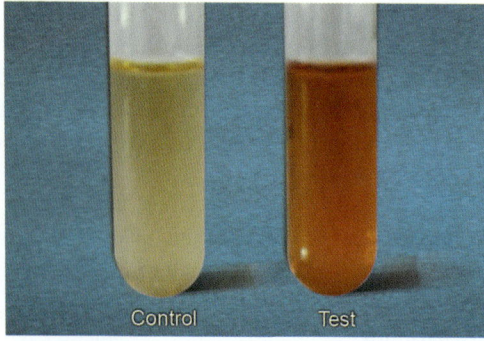

Q. 48
- Identify the test.
- Give the principle and its use in the laboratory.

Ans. 48
- Seliwanoff's test.
- **Principle:** A dehydration reaction due to the hydroxyl groups of the sugar. Concentrated HCl being weaker than concentrated sulfuric acid, dehydrate ketoses (e.g. fructose) more readily than aldoses to form hydroxymethyl furfural which then condenses with resorcinol to form a red colored complex.

Application of the test: Used to differentiate ketoses from aldoses

Spotter-51

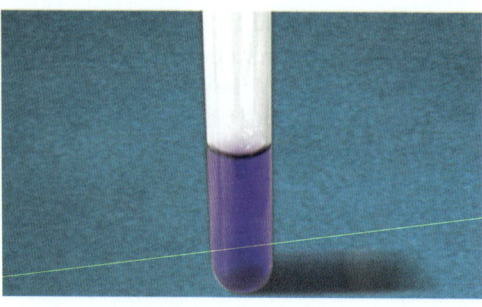

Q. 49
- Identify the test.
- Give the principle and its use in the laboratory.

Ans. 49
- Biuret test.
- **Principle:** The biuret test is given by those substances consist of molecules containing two carbamyl groups ($-CONH_2$) joined either directly or by a single nitrogen or carbon atom. The violet color is due to the formation of a copper coordination complex.

Application of the test: Used as a general identification test of peptides and proteins.

Spotter- 52

Q. 50
- Identify the test.
- Explain the principle and mention its application.

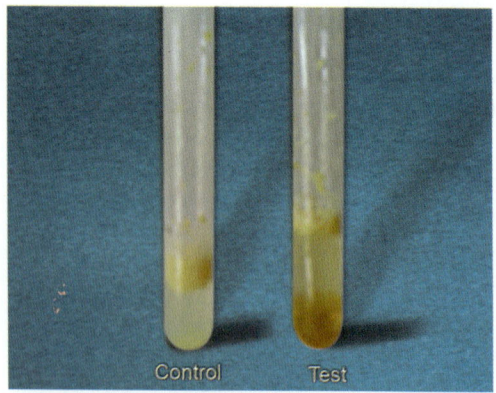

Ans. 50
- Xanthoproteic test → Detect amino acids containing benzene ring, e.g. Tryptophan, tyrosine.
- **Principle:** Addition of nitric acid causes denaturation of proteins to form white precipitate. The yellow color is due to

nitration of phenyl group of amino acids, tryptophan and tyrosine. Addition of alkali increases the ionization of compounds hence the color deepens to get final orange color.

Application of the test: For the detection of benzene group containing amino acids in the test samples.

Spotter- 53

Q. 51
- Identify the test.
- Give the principle and its use in the laboratory.

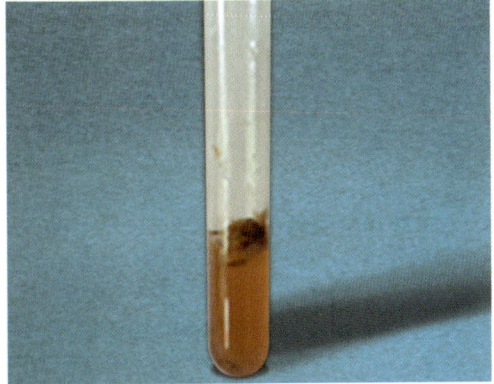

Ans. 51
- Millon's test.
- **Principle**: The protein precipitated by mercuric sulfate in acidic medium to form mercury-protein complex (metalloprotein complex). Nitrous acid is formed by the reaction between sodium nitrite and sulfuric acid. This nitrous acid causes nitration of phenolic groups of tyrosine. Warming enhances nitration process and intensifies the color to give reddish color.

Application of the test: Used to detect the presence of tyrosine in protein solutions or amino acid mixtures.

Spotter- 54

Q. 52
- Identify the test.
- Give the principle and its use in the laboratory.

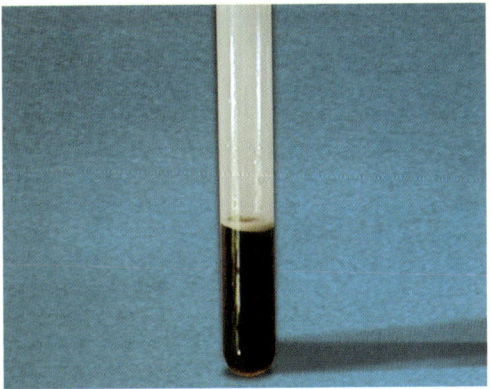

Ans. 52
- Sulfur test.
- **Principle:** Upon boiling with strong alkali the organic sulfur in the cystine and cysteine is converted into inorganic sulfide (here Na_2S). The sodium sulfide react with lead acetate to form black lead sulfide (PbS) and solution turns **brownish black**.

Application of the test: Used to detect the presence of sulfur containing amino acids—cysteine and derived amino acid cystine. It cannot detect the sulfur containing amino acid methionine due to the presence of thioether bond.

Spotter-55

Q. 53
- Identify the test.
- Give the principle and its use in the laboratory.

Ans. 53
- Sakaguchi's test.
- **Principle:** Molisch reagent used in this test contains α-naphthol in alcohol. Sodium hydroxide provides alkaline pH. At the alkaline pH guanidino group of arginine combines with α-naphthol to form bright red color.

Application of the test: Used to detect the presence of the semi-essential amino acid arginine.

Spotter-56

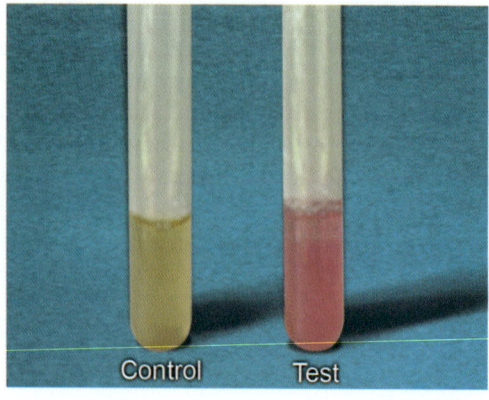

Q. 54
- Identify the test.
- Give the principle and its use in the laboratory.

Ans. 54
- Specific urease test.
- **Principle:** Urease decomposes urea to ammonium carbonate. Ammonium carbonate being basic, raises the pH. Phenol red used in this test will show pink to red color at the basic pH.

$$\text{Urea} \xrightarrow[\text{Urease}]{H_2O} \text{Ammonium carbonate}$$

Application of the test: Used for the identification of urea.

Spotter-57

Q. 55
- Identify the test.
- Give the principle and its use in the laboratory.

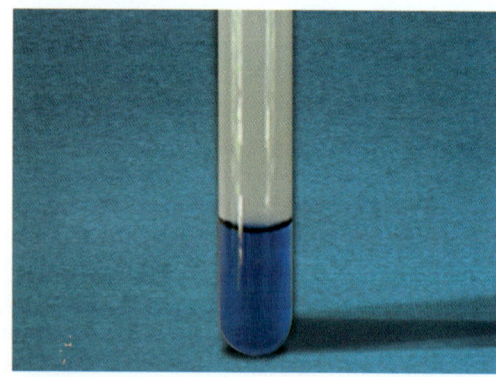

Ans. 55
- Benedict's uric acid test.
- **Principle:** Uric acid reduces phosphotungstic acid to tungsten blue in alkaline medium.

Application of the test: Used for the identification of uric acid.

Spotter-58

Q. 56
- Identify the test.
- Give the principle and its use in the laboratory.

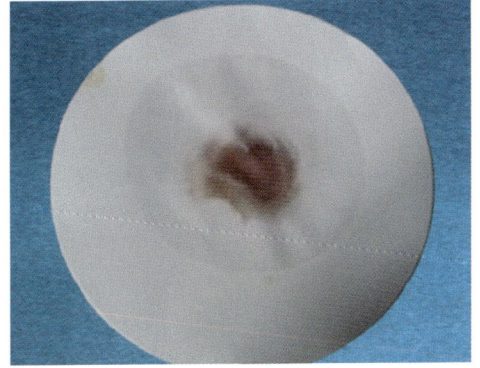

Ans. 56
- Schiff's test.
- **Principle:** Uric acid reduces silver nitrate to metallic silver in alkaline medium.

Application of the test: Used for the identification of uric acid.

Spotter-59

Q. 57
- Identify the test.
- Give the principle and its use in the laboratory.

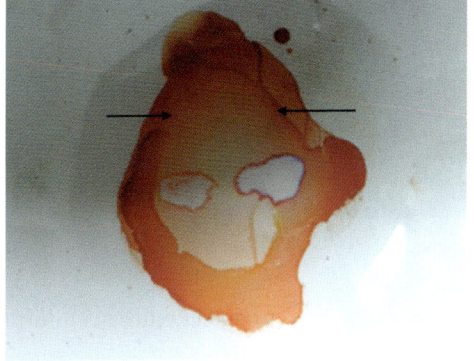

Ans. 57
- Murexide test.
- **Principle:** In this reaction, uric acid is oxidized to dialuric acid and alloxan which condense to form alloxantin. The alloxantin so formed reacts with ammonium hydroxide to form ammonium purpurate or murexide which is purplish red in color. With potassium hydroxide, a purplish violet color is produced due to the formation of potassium salt of ammonium purpurate.

Application of the test: Used for the identification of uric acid.

Spotter-60

Q. 58
- Identify the test.
- Give the principle.
- What is it used for?

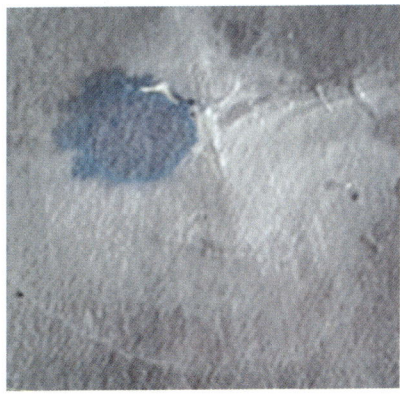

Ans. 58
- Modified Fouchet's test.
- **Principle:** The bile pigment present in the sample is adsorbed on to the precipitate of $BaSO_4$.
  When Fouchet's reagent (Ferric chloride in trichloroacetic acid) is added on to the precipitate, the $FeCl_3$ oxidizes bilirubin to bluish green biliverdin and Fe^{3+} is converted to Fe^{2+}.

- **Application of the test:** Used to detect bilirubin in urine (conjugated bilirubin excreted in urine in obstructive jaundice and hepatic jaundice).

Spotter- 61

Q. 59
- Identify the test.
- Give the principle and its use in the laboratory.

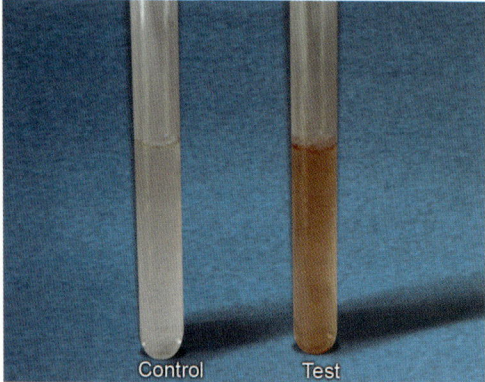

Ans. 59
- Ehrlich's test
  - Interpretation of the test:
    - No red color: Urobilinogen absent
    - Faint pink color: Urobilinogen present in normal amounts
- **Principle**: Urobilinogen forms a colored adduct with Para-dimethyl-aminobenzaldehyde.

Application of the test: Used for the identification of urobilinogen in urine. It is faintly positive in normal urine and strongly positive in hemolytic jaundice.

Spotter- 62

Q. 60
- Identify the test.
- Give the principle and its use in the laboratory.

Ans. 60
- Jaffe test.
- **Principle**: Creatinine forms red colored creatinine picrate with picric acid in alkaline medium. **Application of the test:** Used for the identification and estimation of creatinine in biological fluids.

Spotter- 63

Q. 61
- Identify the test.
- Give the principle of the test.
- What is its use in the laboratory?

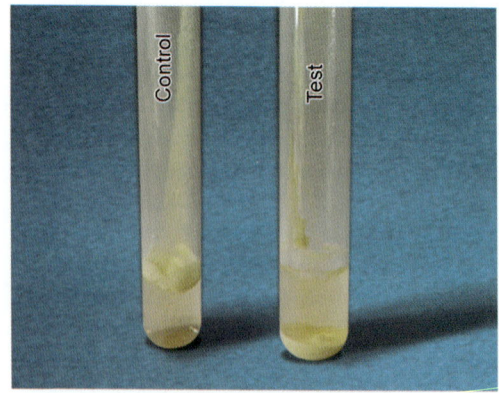

Ans. 61
- Hay's test done to detect bile salts in urine. Test shows positive response.

- **Principle:** Bile salts lower surface tension causing sulfur powder to sink to the bottom.

Application of the test: To diagnose obstructive jaundice in which bile salts are excreted in urine.

NUTRITION

Spotter-64

Q. 62
- Name the limiting amino acids and biological value of rice?
- What is the harm in taking polished rice for long periods especially by persons on poor diet?

Ans. 62
- Limiting amino acids—Lysine, threonine biological value—64%
- Thiamine (vitamin B_1) deficiency causes beriberi.

Spotter-65

Q. 63
- Name the limiting amino acids.
- Biological value of pulse (dal).
- What do you mean by mutual supplementation of proteins?

Ans. 63
- Limiting acid of dal—Methionine
- Biological value—56%
- Mutual supplementation of proteins—Cereals are deficient in the essential amino acids lysine and threonine and pulses deficient in methionine. Generally, the deficiency of these amino acids do not occur due to the habit of taking mixed diet – combination of cereals and pulses, e.g. idli, chapati and dal.

Spotter-66

Q. 64
- What is meant by biological value of a protein?
- Give the biological value of egg.
- What is the harm in taking raw eggs?

Ans. 64
- Biological value = Retained nitrogen/Absorbed nitrogen × 100
- Biological value of egg—90%
- A protein called avidin, which has a high affinity to biotin is present in egg white. When raw egg is consumed, this egg white avidin will bind with dietary biotin tightly in the intestine, thereby blocking the absorption of biotin by intestinal mucosa leading to biotin deficiency.

Avidin being heat labile, cooking of egg will denature avidin and thereby abolish the affinity towards biotin.

Spotter-67

Q. 65
- Name the vitamin that is involved in the collagen synthesis and present in lemon.
- What is the role of this vitamin in the collagen synthesis?
- Name the deficiency manifestation of this vitamin.

Ans. 65
- Ascorbic acid (vitamin C).
- Ascorbic acid is a cofactor of prolyl and lysyl hydroxylase enzymes, involved in the hydroxylation of prolyl and lysyl residues of collagen to form hydroxylysine and hydroxyproline. Hydroxylysine and hydroxyproline are essential for the formation of cross links in the collagen that confer tensile strength to the collagen fibers.
- Scurvy.

Spotter-68

Q. 66
What is the harm in having *Sorghum* as staple diet?

Ans. 66
Leucine content of *Sorghum* is very high. Leucine inhibits the key enzyme QPRTase of nicotinamide nucleotide synthetic pathway, thereby blocking the conversion of niacin to its active form—nicotinamide adenine dinucleotides (NAD and NADP) leading to deficiency manifestations—**pellagra**.

Spotter-69

Q. 67
Mention the cardiovascular health benefits of including fish in the diet.

Ans. 67
Fish oils are rich in omega-3 fatty acids (e.g. Timnodonic acid, cervonic acid). They lower the plasma triglyceride concentration, protect against thrombosis and reduce inflammation.

Spotter-70

Q. 68
- What is the provitamin present in carrot?
- What is the earliest deficiency manifestation of this vitamin?
- Give RDA of this vitamin in adults and children.

Ans. 68
- β carotene is the provitamin of vitamin A.
- Earliest manifestation of vitamin A deficiency: Night blindness.
- RDA of vitamin A: Children – 400–600 µg/day; Adults – 750–1000 µg/day.

Spotter-71

Q. 69
- What is the most abundant type of fatty acid in the sunflower oil?
- What are the benefits of including it in the diet?

Ans. 69
- Sunflower oil contains 63% polyunsaturated fatty acid and it is a good source of essential fatty acids—linoleic acid and linolenic acids.
- Consumption of PUFA helps to reduce serum cholesterol level- by the up regulation of LDL receptors, thereby enhancing the hepatic uptake of LDL and reducing the level of atherogenic lipoprotein fraction LDL, producing an anti-atherogenic effect.

Spotter-72

Q. 70
Give the RDA of the following vitamins in adults:
- Pyridoxine
- Folic acid
- Vitamin C
- Thiamine.

Ans. 70
- *Pyridoxine:* 1–2 mg/day
- *Folic acid:* 200 µg/day; In pregnancy—400 µg/day
- *Vitamin C:* 75–100 mg/day
- *Thiamine:* 1–1.5 mg/day.

Spotter-73

Q. 71
- What is dietary fiber?
- Name any four of them.
- What are the advantages of including high fiber containing items like green leafy vegetables in the diet?

Ans. 71
- *Dietary fiber:* Ingestible carbohydrates in the diet.
- Lignin, Pectins, cellulose and hemicellulose.
- *Advantages:*
  - By improving bowel motility →↓ constipation →↓ in contact with harmful ingredients in the diet with intestinal mucosa →↓ **Ca colon**.
  - By binding with bile acids promote their excretion in feces → promote synthesis of bile acids from cholesterol → help to ↓ serum total cholesterol and ↓ LDL and ↑ HDL fraction—thereby **reduce atherogenic process** in the blood vessels.
  - **Improves glucose tolerance** hence beneficial in diabetes mellitus.

Spotter-74

Q. 72
- Name one branched fatty acid present in the green leafy vegetables.
- How it is metabolized?
- Name the condition in which you will get defective metabolism of such branched fatty acids.

Ans. 72
- Phytanic acid.
- α-Oxidation (to remove the branches) along with β-oxidation.
- Refsum disease.

Spotter-75

Q. 73
- Name one mineral deficient in milk and the one abundantly present in milk.
- Mention the biological value (BV) of milk.

Ans. 73
- Deficient mineral: Iron
- Abundant mineral: Calcium.
- BV of milk: 84.

SPECTROSCOPY

Spotter-76

Q. 74

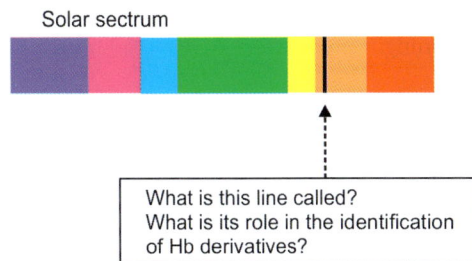

Solar sectrum

What is this line called?
What is its role in the identification of Hb derivatives?

Ans. 74
- D line (the most prominent of Fraunhofer lines)
- D line is used as the reference line in the identification of Hb derivatives by spectroscopic examination.

Spotter-77

Q. 75
A solution of Hb derivative and its absorption spectrum are shown.

Identify the Hb derivative.

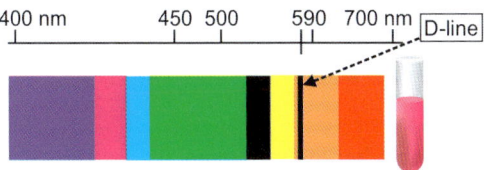

Ans. 75
Deoxy Hb. Identification points:
- *Color of the solution:* Purple.
- *Spectroscopy findings:* A single broad band in the green region. The midpoint of this band corresponds to 565 nm in the green region.

Spotter-78

Q. 76
The solution of Hb derivative and its spectrum are shown.
- Identify the Hb derivative.

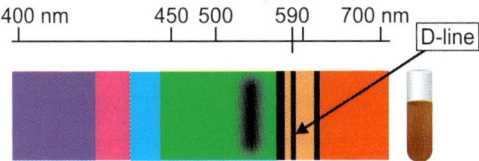

Ans. 76
- **Meth Hb.** Identification points:
  - *Color of the solution:* Reddish brown
  - *Spectroscopy findings:* 3 bands are seen, The characteristic alpha (α) band in the orange region at 630 nm, beta(β) band at 577 nm and a broad gamma (γ) band at 541 nm in the green region.

Spotter-79

Q. 77
The solution of Hb derivative and Hb spectroscopic absorption spectrum are shown.
- Identify the Hb derivative and give its application.

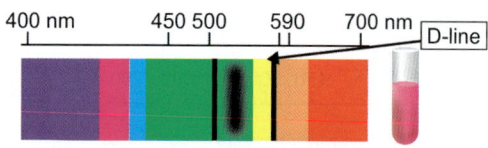

Ans. 77

- **Globin hemochromogen**
  Identification points: 1. Color of the solution: Pink 2. **Spectroscopy findings:** Two characteristic bands are seen—Broad alpha band at 555 nm in the green region and narrow beta band at 525 nm in the green region itself.
  - **Application:** Fetal Hb is relatively resistant to alkali so it will help to distinguish blood rich in HbF from blood containing predominantly adult hemoglobin (Hb A).

Spotter-80

Q. 78

The solution of Hb derivative and Hb absorption spectrum are shown.
- Identify the Hb derivative.

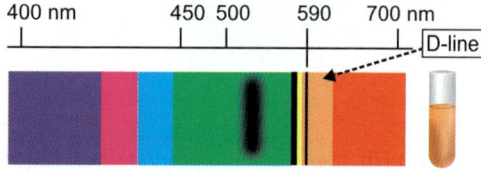

Ans.78

- **Oxy Hb**
  - *Identification points:* Color of the solution: orange red.
  - *Spectroscopy findings:* 2 bands characteristic of oxy Hb are seen. A narrow α band at 577 nm in the yellow region and a broad β band at 541 nm in the green region.

Spotter-81

Q. 79

The solution of Hb derivative and Hb absorption spectrum are shown.
- Identify the Hb derivative and its importance.

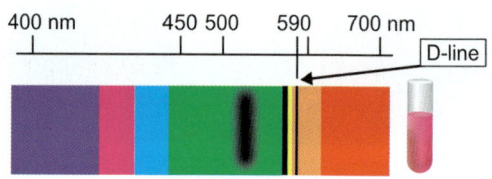

Ans. 79

- **Carboxy Hb**
  - *Identification points:* Color of the solution: Pink.

Spectroscopy findings: Two characteristic bands are seen. α band at 570 nm in the yellow region and β band at 535 nm in the green region. Importance: Identification of carboxy Hb will help in the diagnosis of carbon monoxide poisoning.

CONCEPTUAL QUESTIONS

Spotter-82

Q. 80

- Name this biomolecule.
- What is the consequence produced if this molecule, in the 6th position of β globin chain of hemoglobin is replaced by valine.
- Name one biochemical method to detect such abnormality in an individual.

$$^{-}OOC - CH_2 - CH_2 - CH - COO^{-}$$
$$|$$
$$NH_3$$

Ans. 80
- Glutamic acid.
- Formation of HbS in the body leading to sickle cell anemia.
- Hb electrophoresis.

Spotter-83

Q. 81
Hb has large buffering capacity owing to the presence of the molecule shown in the figure.
- Name the molecule.
- Explain how it confers this property to hemoglobin.

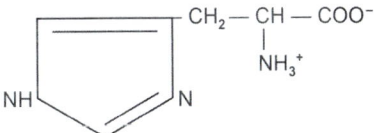

Ans. 81
- Histidine
- pK_a value of ionisable imidazole side chain of histidine is 6.1 which is closer to the physiological pH range of blood. pK_a value of all the other 19 primary amino acids are far away from the physiological pH of blood (i.e. 7.35–7.45). Around 38 histidine residues are present in a molecule of Hb. Hence, the buffering property of Hb is mainly dependent on histidine residues.

Spotter-84

Q. 82
Name the amino acids in order from the amino terminal end.

Single letter abbreviation of a polypeptide is given below:
A V L D F W Q

Ans. 82
Alanine, valine, leucine, aspartic acid, phenylalanine, tryptophan, glutamine (By convention aminoterminal end is placed on the left side).

Spotter-85

Q. 83
- Identify the sugar.
- Name the linkage.
- Mention the name of the intestinal enzyme concerned with its digestion.

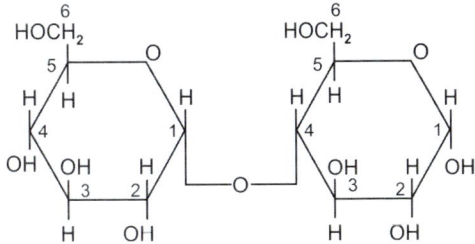

Ans. 83
- Maltose.
- O - α -D – Glucopyranosyl-(1→4) α - D – glucopyranose.
- Maltase.

Spotter-86

Q. 84
- Identify.
- Name the linkage.
- Mention the name of the intestinal enzyme concerned with its digestion.

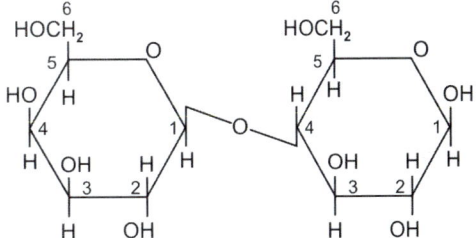

Ans. 84
- Lactose
- O - β -D – Galactopyranosyl-(1→4) β - D – glucopyranose
- Lactase

Spotter-87

Q. 85

- Identify the sugar.
- Name the linkage.
- Mention the name of the intestinal enzyme concerned with its digestion.

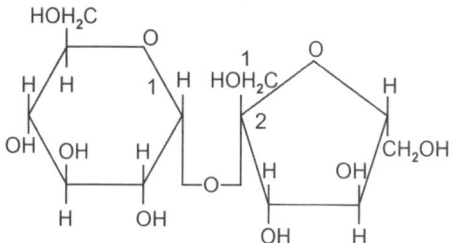

Ans. 85

- Sucrose.
- O - α -D - Glucopyranosyl-(1→2) β - D - fructofuranoside.
- Sucrase.

Spotter-88

Q. 86

- What is the disaccharide present in the mushroom?
- Why do some individuals develop diarrhea, vomiting and abdominal pain upon consuming edible mushrooms?

Ans. 86

- Trehalose.
- Those developing hypersensitivity reactions on taking edible mushrooms is mainly due to **intestinal trehalase deficiency** (that digest trehalose) leading to retention of trehalose which is then fermented by intestinal bacteria to release osmotically active substances that draws water into the lumen leading to diarrhea, vomiting, abdominal distension and abdominal pain.

Spotter-89

Q. 87

- Name the three inhibitors of this reaction.
- Mention the metabolic derangement caused by the deficiency or inhibition of pyruvate dehydrogenase (PDH) enzyme.

Pyruvate dehydrogenase

Pyruvate ⟷ Acetyl CoA

Ans. 87

- Arsenic ions, mercuric ions and deficiency of thiamine.
- Lactic acidosis.

Spotter - 90

Q. 88

- Name the enzyme catalyzing this reaction.
- Name the pathway to which it belongs.
- Give the significance of this enzyme.

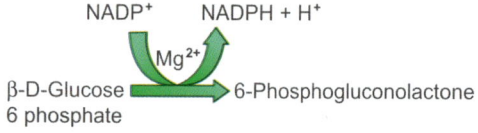

Ans. 88

- Glucose 6 phosphate dehydrogenase (G6PD) catalyze this reaction.
- Hexose monophosphate (HMP) shunt pathway.
- G6PD is the rate limiting enzyme of HMP shunt pathway. Its deficiency produces:
  – Hemolytic anemia in susceptible individuals when they are exposed to oxidants, e.g. oxidant drugs (e.g.

antimalarial primaquin, aspirin, sulfonamides), favabeans
- Neonatal hyperbilirubinemia in neonates.

Spotter -91

Q. 89
- What is the enzyme catalyzing this reaction?
- Name one condition in which its activity is increased.
- Mention one important consequence of that.

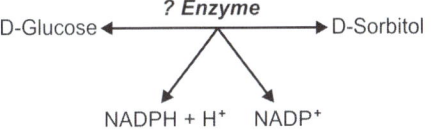

Ans. 89
- Aldose reductase.
- Diabetes mellitus.
- Diabetic cataract.

Spotter -92

Q. 90
- Name the enzymes catalyzing reaction (1) and reaction (2).
- What are the related inborn errors of metabolism?

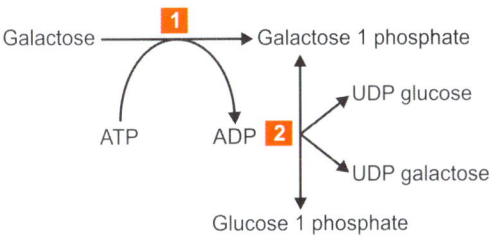

Ans. 90
- Galactokinase catalyze reaction (1) and galactose 1 phosphate uridyl transferase reaction (2).
- Galactosemia.

Spotter -93

Q. 91
- Name the enzymes catalyzing reaction (1) and reaction (2).
- Name the deficiency manifestations of these enzymes.

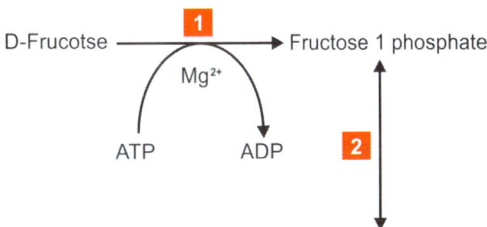

Ans. 91
- Fructokinase catalyze reaction (1) and aldolase B the reaction (2).
- Fructokinase deficiency → essential fructosuria and aldolase B deficiency → hereditary fructose intolerance.

Spotter -94

Q. 92
- Name the enzyme catalyzing the reaction.
- Cite the pathway to which this reaction fit in as a rate limiting step.
- Name two endogenous substances and one drug that inhibit this enzyme and state the purpose of administration of that drug.

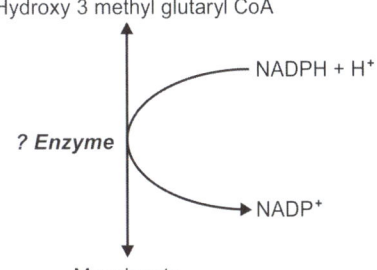

Ans. 92
- HMG CoA reductase.
- Cholesterol biosynthetic pathway.
- Endogenous substances—Cholesterol, bile acids; Drug: Simvastatin coming under statin group of drugs used for lowering serum cholesterol.

Spotter -95

Q. 93
- Name the enzyme catalyzing the rate limiting step of bile acid synthesis.
- Mention two factors inhibiting the activity of this enzyme.
- Name two primary bile acids and two secondary bile acids.

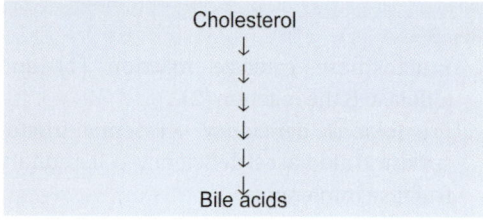

Ans. 93
- Enzyme catalyzing the rate limiting step: 7 - α hydroxylase.
- Inhibiting factors—Vitamin C deficiency and bile acids.
- Primary bile acids: Taurocholic acid and glycocholic acid; Secondary bile acids—Deoxycholic acid and lithocholic acid.

Spotter -96

Q. 94
- Name the products of this reaction.
- Mention one factor each that activate and inhibit this reaction.

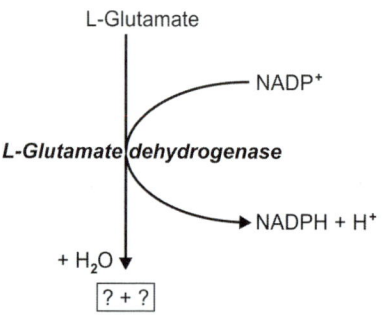

Ans. 94
- Products - α -Ketoglutarate (α - KG) and ammonia.
- Activating factor: ADP; Inhibiting factor: GTP.

Spotter -97

Q. 95

Clinical condition	Type of bilirubinemia	Defect
Neonatal physiological jaundice	a.?	b.?
Crigler-Najjar syndrome type I	c.?	d.?
Gilbert syndrome	e?	f?
Dubin Johnson syndrome	g?	h?

Go through the table and answer accordingly.

Ans. 95
a. Unconjugated hyperbilirubinemia
b. Accelerated hemolysis around the time of birth and immature hepatic function (uptake, conjugation and secretion of bilirubin).
c. Unconjugated hyperbilirubinemia.

d. Absence of UDP glucuronyl transferase (UDP-GT) in the hepatocytes.
e. Unconjugated hyperbilirubinemia.
f. Mild deficiency of UDP glucuronyl transferase (UDP-GT) activity in the hepatocytes.
g. Conjugated hyperbilirubinemia.
h. Defective ATP dependent organic anion transporter → impaired hepatic excretion of bilirubin.

Spotter -98

Q. 96
- Name the enzyme catalyzing this reaction.
- What is the importance of this enzyme in the treatment of gout.

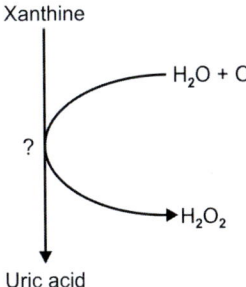

Ans. 96
- Enzyme catalyzing the reaction—Xanthine oxidase.
- The drug allopurinol (a purine analog) inhibits xanthine oxidase → lowering of uric acid level → relief in **gout**.

Spotter -99

Q. 97
- Identify the structure.
- How can it be produced?
- What is its use in the medical field?

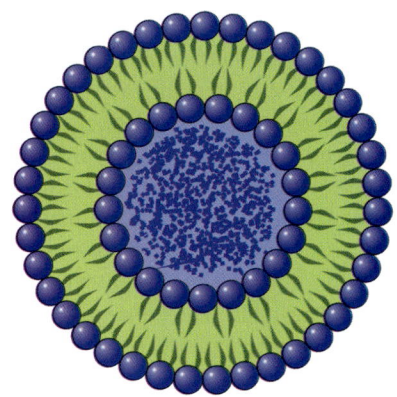

Ans. 97
- Liposome. It consists of spheres of lipid bilayers that enclose an aqueous medium.
- It can be produced by sonicating amphipathic lipids in an aqueous medium.
- a. Liposomes can be loaded with drugs for tissue-specific delivery of drugs in cancer therapy with minimum side effects.
- b. Liposomes can be used as a carrier of genes, in gene therapy.

Spotter -100

Q. 98
- Identify the structure.
- How does it form in the gut?
- What is its significance?

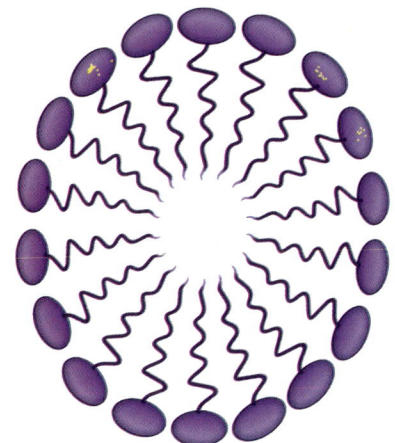

Ans. 98
- Micelle. It is a sphere of lipid monolayer.
- Major lipids of the dietary triglycerides and phospholipids hydrolyze to form amphipathic lipids and are emulsified to micelles in the gut. Micelles are formed when a critical concentration of amphipathic lipids (fatty acids, phospholipids, sphingolipids, bile salts and cholesterol in an aqueous medium) reaches in an aqueous medium.
- Allow absorption of lipids and fat-soluble vitamins from the gut.

Spotter -101

Q. 99
- Identify the compound.
- Name two important compounds derived from it in the body.
- What is the rate limiting enzyme of its synthetic pathway?

Ans. 99
- Cholesterol (3-hydroxy-5, 6 –cholestene).
- Vitamin D and steroid hormones.
- HMG CoA reductase.

Spotter -102

Q. 100
- Identify the virus.
- Name the stages of the disease caused by this virus and most suitable tests for the diagnosis of each stage.

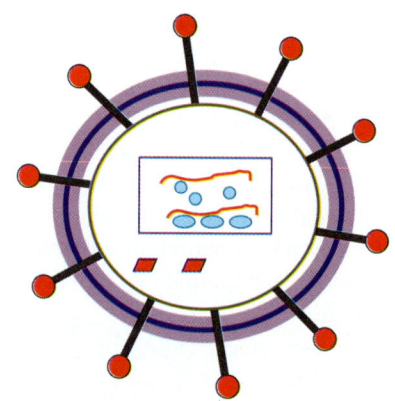

Ans. 100
- Human immunodeficiency virus.
- Stages of HIV infection and suitable tests for diagnosis.

Stages	Tests
Stage 1 – Acute viral syndrome	• PCR for HIV RNA • p 24 antigen assay
Stage 2 – Seropositive stage (asymptomatic)	Antibody detection by • ELISA – detects antibodies against single HIV antigen • Western blot – detects antibodies against 6 – 8 HIV antigens – hence more reliable
Stage 3 – Acquired immunodeficiency syndrome (AIDS)	• PCR for HIV RNA, p 24 antigen assay • CD4 cell count – Low

Spotter -103

Q. 101
- Name the phase of cell cycle in which DNA synthesis occurs.
- Name three important restriction points (check points).
- Mention the check point at which guardian of genome (p53) act to block cell cycle.

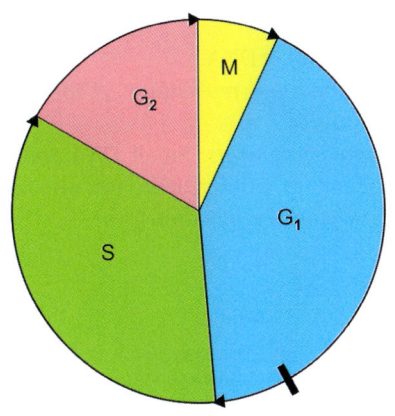

Ans. 101
- S phase.
- 1) G_1-S transition, 2) S phase, 3) G_2-M boundary.
- G_1 - S transition.

Spotter -104

Q. 102
Mention two examples each for the following:
- Tumor suppressor genes
- Oncogenes
- Viruses causing cancer
- Tumor markers
- Chemical carcinogens
- Physical agents causing cancer
- Antimutagens.

Ans. 102
- p53, RB.
- ABL, ERB – B1.
- Epstein-Barr virus (EBV) → Nasopharyngeal cancer hepatitis B virus → hepatoma.
- AFP (alpha fetoprotein) to detect liver cancer, CEA to detect colorectal cancers.
- Afla toxin, Nitroso compounds (e.g. Dimethyl nitrosamine).
- X rays, UV rays.
- Vitamin E, vitamin C.

Spotter -105

Q. 103
- Name three auxiliary proteins of protein folding.
- Name two protein conformational disorders.

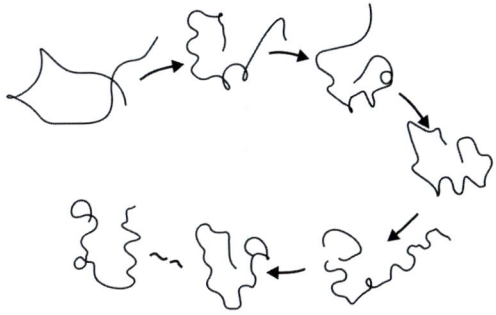

Ans. 103
- 1) Chaperones (e.g. HSP 60, HSP 70, 2) Protein disulfide isomerase, 3) Proline cis trans isomerase.
- Kuru, Creutzfeldt-Jakob disease (CJD).

Spotter -106

Q. 104
- Name the type of gene regulation operating in drug resistance exerted by the anticancer drug methotrexate.
- Name any other two mechanisms of gene regulation.

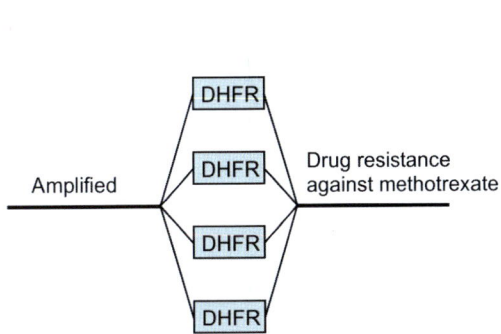

Ans. 104
- Gene amplification.
- Gene rearrangement, alternative RNA splicing.

Spotter -107

Q. 105
- Identify structure and mention the function.
- Name two inhibitors of protein synthesis.
- Name two post-translational modifications.

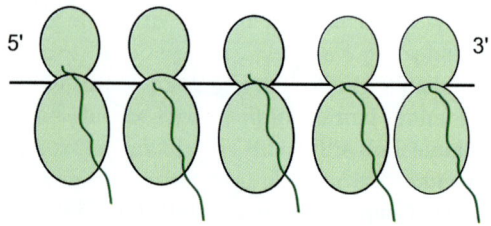

Ans. 105
- Polysomes function: Translation of mRNA → Protein synthesis.
- Tetracycline, chloromycetin.
- Loss of signal sequences, amino terminal and carboxy terminal modifications.

Spotter -108

Q. 106
- Name the process illustrated here.
- Name two inhibitors of this process.
- What are Okazaki fragments?

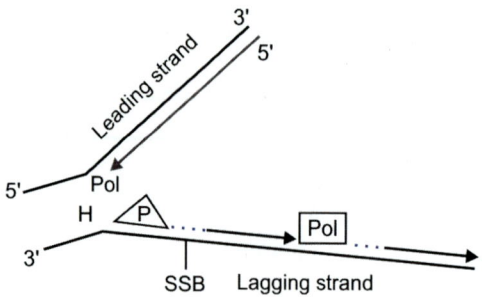

Ans. 106
- Replication of DNA.
- *Inhibitors of replication:* Ciprofloxacin, adriamycin.
- *Okazaki fragments:* Small DNA molecules attached to primer RNA (on the lagging strand side of replication fork).

Spotter -109

Q. 107
- What is the mechanism depicted here?
- Name one disease produced by the deficiency of this mechanism in the body.

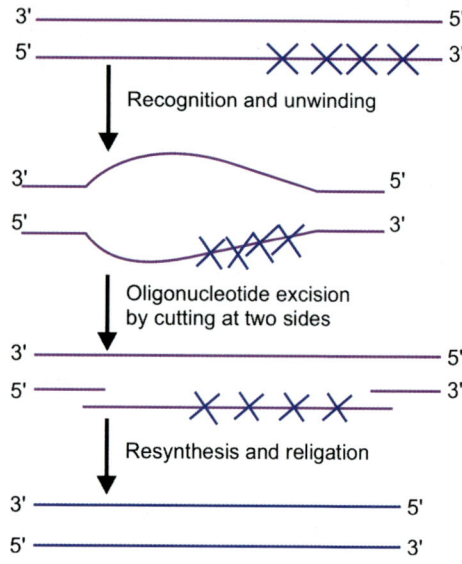

Ans. 107
- Nucleotide excision repair
- Xeroderma pigmentosum.

Spotter -110

Q. 108
- What are the ends of chromosomes called?
- What is the enzyme involved in the replication ends of DNA?
- What is the significance of this enzyme in cancer cells and aging?

Ans. 108
- Telomere.
- Telomerase.
- *Cancer cells:* Telomerase activity increased; Aging: Telomerase activity decreased → senescence.

Spotter -111

Q. 109
- What is the pattern of inheritance illustrated here?
- Name two diseases inherited by this pattern.

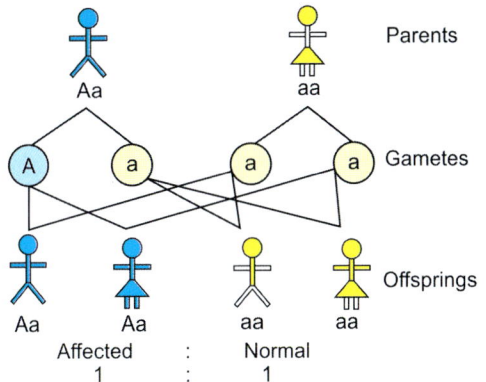

Ans. 109
- Autosomal dominant
(Features: Manifest in the heterozygous state, one parent of an index case affected; offsprings will have 50% chance of getting the disease; both males and females are affected).
- Marfan syndrome, polycystic kidney disease.

Spotter -112

Q. 110
- What is the pattern of inheritance illustrated here?
- Name two diseases inherited by this pattern.

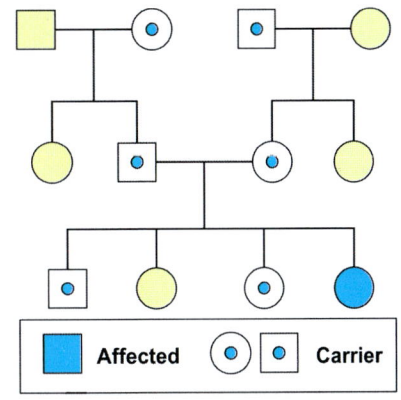

Ans. 110
- Autosomal recessive
(Features: Manifest in the homozygous state, offsprings will have 25% chance of getting the disease and 25% chance of getting the carrier state when both the parents are heterozygous as shown in the illustration).
- Sickle cell anemia and phenylketonuria.

Spotter -113

Q. 111
- Name the pattern of inheritance illustrated here.
- Name two diseases inherited by this pattern.

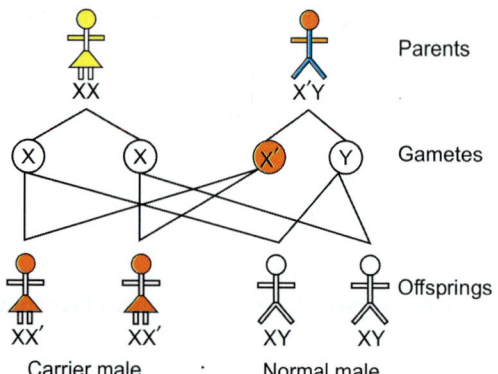

Carrier male : Normal male

Ans. 111
- X-linked recessive.
- G6PD deficiency, Duchenne type pseudo-muscular dystrophy.

Spotter -114

Q. 112
- What is the pattern of inheritance illustrated here?
- Mention three characteristics of this pattern of inheritance.
- Cite one example of a disease for this type of inheritance pattern.

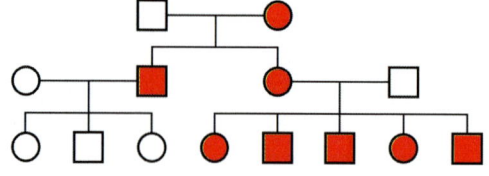

Ans. 112
- Maternal (mitochondrial) inheritance.
- (1) Not inherited according to Mendel's laws, affected mother passes disease to all children but affected father do not

(2) Transmission of abnormal mitochondrial genes ends with each son
(3) Only daughters pass the disease to offspring.
- Leber hereditary optic neuropathy.

Spotter -115

Hb A α chain   58, His   CAU   or   CAC
↓   ↓   ↓   ↓
Hb M (Boston)   58, Tyr   UAU   UAC
α chain

Q. 113
What is the type of mutation?

Ans. 113
Unacceptable missense mutation.

Spotter -116

Q. 114
What is the type of mutation?
Hb A-β chain   6 Glu   GAA   or   GAG
↓   ↓   ↓   ↓
Hb S-β chain   6 Val   GUA   GUG

Ans. 114
Sickle cell anemia/trait: Partially acceptable missense mutation.

Spotter -117

Q. 115
What is the type of mutation?
Hb A, β Chain   61 Lys   AAA or AAG
↓   ↓   ↓
Hb Hikari, β chain   61 Asn   AAU or AAC

Ans. 115
Acceptable missense mutation.

SECTION 5

Objective Structured Practical Examination

CHAPTER
25

Objective Structured Practical Examination: Model Questions

GENERAL GUIDELINES

- OSPE questions prepared for experiments should be observable and structured.
- In addition to experiment, one or two questions can be asked by the observer at the station to derive the concealed concepts related to the experiment, e.g. inference/clinical correlations/reference range, etc.
- A few model OSPE questions are furnished below along with check list for structured observations.
- A set of common laboratory reagents and laboratory ware should be arranged at the station. So that students can select the required reagents according to the OSPE questions.
- Time should be fixed depending on the type of OSPE given
- Advantages of OSPE:
    - If correctly done, full marks can be scored.
    - Besides, practicing OSPE questions will sharpen the skills.

Q. 1
A solution suspected to contain carbohydrate solution is provided. Do a test to detect its presence.

Steps	Observation Points	Marks
1.	**Selection of general test for detecting a carbohydrate:** Molisch's test.	1
2.	Take 3 mL of sugar solution in a test tube, add two drops of Molisch reagent. Mix thoroughly.	1
3.	Add 3 mL of concentrated sulfuric acid along the sides of the test tube by slightly inclining the tube, so that a violet ring forms.	1
	Questions	
1.	What are the ingredients of Molisch reagent?	1
2.	What is the principle of the test?	1

Ans. 1

1. *Ingredients:* Alpha (α) naphthol in alcohol.
2. *Principle:* Concentrated acid dehydrates the sugar to form furfural (in the case of pentoses) or furfural derivatives (hexoses and heptoses) which then condenses with α-naphthol to give a **violet** colored complex.

Q. 2

Glucose solution is provided in a numbered beaker/test tube. Show that it is an aldose by doing one test.

Steps	Observation Points	Marks
1.	**Selection of a test to show glucose is an aldose:** Rapid furfural test.	1
2.	To 2 mL of concentrated HCl, add 8 drops of sugar solution and 1–2 drops of Molisch reagent.	1
3.	Mix well and heat just to boil. No formation of violet color.	1
	Questions	
1.	Name one more test useful for demonstrating glucose is an aldose.	1
2.	What is the principle of the test done?	1

Ans. 2

1. Seliwanoff's test
2. *Principle of rapid furfural test:* Concentrated HCl being weaker than concentrated sulfuric acid, dehydrate ketoses (e.g. fructose) more readily than aldoses to form hydroxymethyl furfural, which then condenses with α-naphthol to form a violet colored complex.

Q. 3

Glucose solution is provided in the numbered beaker. Give the approximate concentration of the solution by doing a qualitative test.

Steps	Observation Points	Marks
1.	**Selection of the test:** Benedict's test.	1/2
2.	Take 5 mL of Benedict's reagent in a test tube and add exactly 8 drops (0.5 mL) of the sugar solution. Mix well. Boil for 2 minutes or keep in a boiling water bath for 3 minutes.	1
3.	Allow the contents to cool spontaneously.	1/2
4.	Derive the approximate concentration of glucose solution indicated by the color of the precipitate.	1
	Questions	
1.	What is the principle of Benedict's test?	1
2.	Name two non-carbohydrate substances answered by Benedict's test.	1

Ans. 3

1. *Principle:* Carbohydrates with a free aldehyde or keto group have the ability to reduce various metallic ions. In this test cupric ions are reduced to cuprous ions by the enediols formed from sugars in the alkaline medium of Benedict's reagent.
2. Ascorbic acid and homogentisic acid.

Q. 4

Fructose solution is supplied. Demonstrate that fructose is a ketosugar.

Steps	Observation Points	Marks
1.	Selection of a test to show fructose is a ketose: Seliwanoff's test.	1
2.	To 3 mL of Seliwanoff's reagent in a test tube, add 5 drops of fructose solution.	1
3.	Heat the contents to **just boiling**. Forms a red color within 30 seconds.	1
	Questions	
1.	Name two more tests useful for demonstrating fructose is a ketose.	1
2.	What is the principle of the test done?	1

Ans. 4

1. Rapid furfural test and Seliwanoff's test
2. *Principle:* A dehydration reaction due to the hydroxyl groups of the sugar. Seliwanoff's reagent is resorcinol in dilute hydrochloric acid. Ketoses (e.g. fructose) are more readily dehydrated by HCl than the aldoses to form hydroxymethyl furfural which then condenses with resorcinol of Seliowanoff's reagent to form a **red colored** complex.

Q. 5

A carbohydrate solution is supplied. Demonstrate that the solution contains a disaccharide.

Steps	Observation Points	Marks
1.	Selection of a test to show the carbohydrate present in the solution is a disaccharide: Barfoed's test.	1
2.	Take 5 mL of Barfoed's reagent in a test tube and add 0.5 mL of sugar solution. Mix well.	1
3.	Keep in a boiling water bath for **2 minutes**. Negative response (no red precipitate) of the test indicates that the carbohydrate supplied is a disaccharide.	1
	Questions	
1.	What is principle of the test done?	1
2.	Name two reducing disaccharides.	1

Ans. 5

1. *Principle:* It is a reduction test. Reducing property owes to the carbonyl group (aldehyde or keto group) of sugars. Barfoed's reagent is copper acetate in acetic acid. The cupric ions are reduced to cuprous ions which forms red cuprous oxide by enediol forms of sugars. Monosaccharides enolize within 2 minutes in weak acidic medium provided by Barfoed's reagent where as disaccharides do not. Hence, disaccharides give a negative response.
2. Lactose and Maltose.

Q. 6

Do a test to show that the given solution contains sucrose.

Steps	Observation Points	Marks
1.	**Selection of a test to demonstrate sucrose:** Specific sucrose test.	1
2.	Take 3 mL of sucrose solution and add 1 drop of thymol blue indicator and one or two drops of dilute HCl to obtain a pink color. Divide it into two equal parts.	1
3.	Boil one part for 1 minute and the other part is kept as control. Neutralize both parts by adding 20% sodium carbonate drop by drop until a blue color develops.	1
	Do Benedict's test on hydrolysate and on control.	1
	Unboiled sucrose solution will not give a positive response to Benedict's test whereas boiled portion gives a positive response.	1
	Questions	
1.	What is the purpose of boiling?	1
2.	Give acidic and basic pH ranges and corresponding color ranges of thymol blue.	1

Ans. 6

1. Boiling enhances acid hydrolysis of sucrose to form glucose and fructose.
2. Thymol blue indicator contains two components that work at acid range (pH range 1.2–2.8; color change– red to yellow) and at alkaline range (pH range 8.0–9.6; color change–yellow to blue).

Q. 7

Name the homopolysaccharide present in rice and do a test to demonstrate its presence in the given solution.

Steps	Observation Points	Marks
1.	**Homopolysaccharide present in rice:** Starch **Test to demonstrate starch:** Iodine test	½ + ½
2.	Take 2–3 mL of given solution add 2 drops of dilute (0.05 N) iodine solution	1
3.	Deep blue color appears on adding iodine solution which then disappears on heating and then reappears on cooling.	1
	Questions	
1.	What is the inference?	1
2.	Name one example each for a homopolysaccharide and a heteropolysaccharide present in the body?	1

Ans. 7

1. *Inference of iodine test:* Starch forms a adsorption complex with iodine to give a blue color. The blue color disappears on heating due to the breaking of the Iodine starch adsorption complex and appears on cooling due to reformation of the adsorption complex.
2. Homopolysaccharide, e.g. glycogen; heteropolysaccharide, e.g. glycosaminoglycans (mucopolysaccharides) like heparin sulfate, chondroitin sulfate, dermatan sulfate

Q. 8

Demonstrate protein in the given solution.

Steps	Observation Points	Marks
1.	**Test to demonstrate protein:** Biuret test.	1
2.	To 2–3 mL of protein solution in a test tube, add an equal volume of 10% sodium hydroxide solution, mix thoroughly.	1
3.	Then add a 0.5% copper sulfate solution drop by drop, mixing between drops until a purplish violet color is obtained.	1
	Questions	
1.	What is the principle of the test done?	1
2.	Name the most abundant protein in plasma.	1

Ans. 8

1. *Principle:* The biuret test is given by molecules containing two carbamyl groups (– $CONH_2$) joined either directly or by a single nitrogen or carbon atom. The purplish violet color is due to the formation of a copper coordination complex with peptide bond nitrogen atoms.
2. Albumin.

Q. 9

Do Heller's test demonstrate protein in the urine sample supplied.

Steps	Observation Points	Marks
1.	Take 2 mL of protein solution in a test tube. Add 2 mL of concentrated HNO_3 along the sides of the test tube slowly.	1
2.	A white ring forms at the junction of two liquids.	1
	Questions	
1.	What is the principle of the test done?	1
2.	Name another test to detect presence of proteins in urine.	1

Ans. 9

1. *Principle:* Stratification of albumin solution over strong mineral acid causes denaturation and precipitation of albumin at the point of contact, that is at the junction between two layers.
2. Heat coagulation test.

Q. 10

Demonstrate the amino acid Tyrosine, in the given protein solution.

Steps	Observation Points	Marks
1.	Take 2 mL of protein solution in a test tube and add 2 mL of 10% mercuric sulfate ($HgSO_4$) in 10% sulfuric acid (Millon's reagent).	1
2.	Boil for 30 seconds. A precipitate may form at this stage. Cool and add a few drops of 1% $NaNO_2$ and gently warm.	1
3.	Yellow precipitate forms and turns red on heating.	
	Questions	
1.	Name two biologically important compounds derived from Tyrosine in the body.	1
2.	Name the amino acid from which Tyrosine is formed in the body.	1

Ans. 10

1. Epinephrine and thyroid hormones.
2. Phenyl alanine.

Q. 11

Demonstrate the amino acid Tryptophan, in the given solution.

Steps	Observation Points	Marks
1.	Take 2 mL of test solution add 2 drops of 1/500 formaldehyde (HCHO) and 1 drop of 10% mercuric sulfate in sulfuric acid (Millon's reagent). Mix well.	1
2.	Add 2 mL of concentrated sulfuric acid through the sides of the test tube.	1
3.	A **purple** ring develops at the junction of two layers.	1
	Questions	
1.	Name one vitamin derived from Tryptophan in the body.	1
2.	What is the one letter abbreviation for Tryptophan?	1

Ans. 11

1. Niacin.
2. W.

Q. 12

Demonstrate the amino acid Arginine, in the given solution.

Steps	Observation Points	Marks
1.	Add 5 drops of 5% sodium hydroxide to 5 mL of protein solution. Shake well. Add 2–4 drops of Molisch's reagent.	1
2.	Add 2 mL freshly prepared bromine water.	1
3.	A **bright red** color develops.	1
	Questions	
1.	Name the characteristic side group of this amino acid.	1
2.	Name any one biologically important substance formed from it.	1

Ans. 12

1. Guanidino group
2. Nitric oxide (NO) that serves as a neurotransmitter, smooth muscle relaxant, and vasodilator.

Q. 13

Demonstrate the amino acid cysteine, in the given solution.

Steps	Observation Points	Marks
1.	To 3 mL of protein solution add 3 mL of 40% NaOH and boil for 3 minutes.	1
2.	Cool, add 1 mL of lead acetate solution.	1
3.	Solution turns dark brown.	1
	Questions	
1.	What is the principle of this test?	1
2.	Name the other primary sulfur containing amino acid and its active form in the body.	1

Ans. 13

1. *Principle of sulfur test:* Upon boiling with strong alkali the organic sulfur in the cysteine is converted into sulfide (here Na_2S). The sodium sulfide reacts with lead acetate to form black lead sulfide (PbS) and hence the solution turns brownish black.
2. Methionine; Active form of methionine is S–Adenosyl Methionine (SAM).

Q. 14

Demonstrate the amino acid Histidine, in the given solution.

Steps	Observation Points	Marks
1.	**Pauly's test:** Take 0.5 mL of 0.5% sulfanilic acid in HCl in a test tube and add an equal volume of 0.5% freshly prepared sodium nitrite.	1
2.	Allow to stand for 1 minute and add 1 mL of protein solution. Mix well and add 1 mL of 10% Na_2CO_3 to make the solution alkaline.	1
3.	**Cherry red** color may be observed.	1
	Questions	
1.	Name one compound derived from Histidine in the body?	1
2.	What is the pKa of Histidine? Mention its importance in connection with acid-base balance.	½ + ½

Ans. 14

1. Histamine.
2. pKa of imidazole group of Histidine is 6 which is nearer to blood pH and hence the buffering action of proteins (e.g. serum albumin and hemoglobin) owes to the presence of histidine.

Q. 15

Demonstrate presence of aromatic amino acids in the albumin solution supplied.

Steps	Observation Points	Marks
1.	**Xanthoproteic Reaction:** Add 1 mL of concentrated nitric acid to 2–3 mL of test protein solution taken in a test tube. Heat to boil. A white precipitate forms on adding nitric acid, on heating turns yellow and then dissolves to impart yellow color to the solution.	1
2.	Cool and pour half of the solution into another tube. One tube is kept as control and the other as test.	1
3.	To one tube add 40% NaOH in drops. Upon adding alkali the color deepens to attain **orange** color.	1
	Question	
1.	Name three aromatic amino acids?	1

Ans. 15

1. Phenyl alanine, Tyrosine and Tryptophan.

Q. 16

Do heat and acetic acid with the albumin solution and write down the principle of the test.

Steps	Observation Points	Marks
1.	Take a test tube and fill albumin solution up to two-thirds. Heat the upper one-third of the column of protein solution. Note whether the precipitate formed earlier (if any) becomes intensified or appears upon adding acetic acid.	1
2.	Irrespective of the presence or absence of the appearance of the precipitate, add 2% acetic acid drop by drop.	1
3.	Note whether the precipitate formed earlier (if any) becomes intensified or appears upon adding acetic acid.	1
	Question	
1.	What is the purpose of adding acetic acid?	1

Ans. 16

1. Addition of acetic acid helps
   - To lower the pH to isoelectric point of albumin to maximize the precipitation.
   - To dissolve the precipitate due to phosphates.

Q. 17

Demonstrate the presence of phosphorus in the casein solution supplied.

Steps	Observation Points	Marks
1.	**Neumann's test (detect organic phosphorus):** To 5 mL of casein solution in a test tube, add 0.5 mL of 40% NaOH. Heat for one minute and cool it keeping in a rack.	1
2.	Add 0.5 mL of concentrated nitric acid. Add 1 mL of saturated ammonium molybdate solution. Canary yellow precipitate.	1
	Questions	
1.	What is this canary yellow precipitate due to?	1
2.	Name one food item that contains casein.	1

Ans. 17

1. Canary yellow precipitate is Ammonium phosphomolybdate.
2. Milk contains casein.

Q. 18

Prove that the given protein solution is casein and not albumin by doing one test.

Steps	Observation Points	Marks
1.	**Half saturation test:** To 5 mL of casein solution in a test tube add an equal volume of saturated ammonium sulfate solution. Shake vigorously for 2 minutes. Keep it for 5 more minutes.	1
2.	Filter and collect the filtrate.	1
	Perform biuret test with the filtrate. To 2 mL of the above filtrate taken in a test tube add 2 mL 40% NaOH and 1% $CuSO_4$ drop by drop. If positive, the given solution contains albumin and if negative indicates complete precipitation of casein by half saturation.	1
	Questions	
1.	What are the other two tests useful to confirm the presence of casein?	1
2.	Name one food item that contains casein.	1

Ans. 18

1. Positive Neumann's test and weakly positive Sulfur test.
2. Milk contains casein.

Q. 19

Prove that the given substance is fat by doing one test.

Steps	Observation Points	Marks
1.	**Grease spot test:** Place a drop substance/pinch on a piece of ordinary writing paper.	1
2.	A translucent spot develops.	1
	Question	
1.	What are the other two tests useful in this context?	1

Ans. 19

1. Solubility test and Acrolein test.

Q. 20

Two specimens of oils are supplied. Identify them as saturated or unsaturated type.

Steps	Observation Points	Marks
1.	**Halogenation test:** Take two test tubes and mark A and B respectively. • Add 5 mL of chloroform to both tubes. • Add 8 drops of one type of oil to tube A and add 8 drops of the type of oil to tube B and shake well.	1
2.	Add a few drops of fresh bromine water in both tubes and shake well.	1
3.	The tube in which orange yellow color of bromine water vanishes indicates the presence of unsaturated fatty acid.	1
4.	The tube in which orange yellow color of bromine water persists indicates the presence of saturated fatty acid.	1
	Question	
1.	Name one saturated and one monounsaturated fatty acid.	1

Ans 20

1. Palmitic acid and Palmitoleic acid.

Q. 21

Identify the crystal by microscopy and draw the shape of it.

Step	Observation Point	Marks
1.	Cholesterol crystal– rhombic crystals notched at one corner.	1
	Question	
1.	Name two chemical tests to identify it.	1

Ans. 21

1. Salkowski's reaction (H_2SO_4 test), Liebermann-Burchard reaction (acetic anhydride sulfuric acid test).

Q. 22

Do one test to detect urea in the given solution.

Step	Observation Point	Marks
1.	**Alkaline hypobromite test:** To 3 mL of urea solution in a test tube add a few drops of alkaline hypobromite solution (3 mL concentrated NaOH + 2 mL bromine water).	1
	Brisk effervescence.	1
	Question	
1.	What is this brisk effervescence due to?	2

Ans. 22

1. Brisk effervescence is due to nitrogen gas.

Q. 23

Show that the given solution contains creatinine.

Step	Observation Point	Marks
1.	**Jaffe's test (picric acid reaction):** To 5 mL of creatinine solution in a tube, add 1 mL of 1% picric acid and 10 drops of 10% NaOH. Shake well and keep it for a few minutes.	1
	An orange red color forms.	1
	Question	
1.	What is the principle of this test?	2

Ans. 23

1. *Principle:* Creatinine forms creatinine picrate in alkaline medium which is orange red in color.

Q. 24

Demonstrate the presence of uric acid in the given solution.

Steps	Observation Points	Marks
1.	**Benedict's uric acid test:** To 5 mL of uric acid solution in a tube, add 1 mL of 1% Na_2CO_3 and a few drops of Benedict's uric acid reagent.	1
2.	Intense blue color.	1
	Questions	
1.	What is the principle of this test?	1
2.	Name one more test to detect uric acid.	1

Ans. 24

1. Uric acid reduces phosphotungstic acid to tungsten blue in alkaline medium.
2. Schiff's test.

Q. 25

Check the pH of urine and specific gravity.

Steps	Observation Points	Marks
1.	Use red and blue pH paper to check pH.	1
2.	**To check specific gravity:** Use urinometer Urine taken in a urine glass, urinometer made to float in it, read the specific gravity from the scale—the point corresponding to the upper meniscus.	3
	Questions	
1.	What is the normal pH range of urine?	1
2.	What is the normal range of specific gravity of urine?	1

Ans. 25

1. *Urine pH:* 4.6–8.
2. *Urine specific gravity:* 1.015–1.025.

Q. 26

Demonstrate the presence of chloride in the normal urine supplied.

Steps	Observation Points	Marks
1.	Acidify 2 mL of urine with 2 drops of concentrated HNO_3 and add 2 mL of silver nitrate solution.	1
2.	White precipitate.	1
	Questions	
1.	Mention the normal excretion rate?	1
2.	Mention the condition in which its concentration in urine becomes high?	1

Ans. 26

1. *Urinary excretion rate of chloride:* 10–15 g /day.
2. Addison's disease in which there is aldosterone deficiency so that reabsorption sodium and chloride are defective → increased excretion in urine.

Q. 27

Make a report of five physical properties of the sample of urine provided.

Steps	Observation Points	Marks
1.	1. Appearance 2. Odor 3. Color 4. pH 5. Specific gravity.	½ + ½ + ½ + ½ + 1
	Question	
1.	Mention one condition each for the different colors of urine written below: i. Deep yellow ii. Red color iii. Black color.	3

Ans. 27

1. i. Jaundice; ii. Hematuria; iii. Alkaptonuria

Q. 28

Demonstrate the presence of inorganic sulfate in the normal urine supplied.

Steps	Observation Points	Marks
1.	Acidify 3 mL of urine with 2–4 drops of concentrated HCl and add 1 mL of barium chloride solution.	1
2.	White precipitate.	1
	Question	
1.	Mention the normal excretion rate and the source of it.	1 + 1

Ans. 28

1. *Urinary excretion rate of sulfate:* 0.8–1.0 g /day.
   Source: Sulfur containing amino acids (cysteine, methionine).

Q. 29

Demonstrate the presence of calcium in the normal urine supplied.

Steps	Observation Points	Marks
1.	Take 10–12 mL of urine in a boiling tube add 3 mL of strong ammonia solution and boil till white precipitates of calcium and magnesium are formed.	1
2.	Filter through a filter paper placed in a funnel placed over a test tube. Wash the precipitate collected in the filter paper by just pouring a few mL of water through the filter paper.	1
	Take the funnel with the filter paper in situ, and place it over another test tube. Add 3 mL hot acetic acid through the filter paper placed over the test tube. Hot acetic acid with dissolved precipitate is collected in the test tube underneath. Divide it into 2 parts.	
3.	**To detect calcium:** To one part add 1 mL of potassium oxalate → White precipitate.	1
	Question	
1.	Mention the normal excretion rate and the source of it.	1 + 1

Ans. 29.

1. *Urinary excretion rate of calcium:* 0.1–0.3 g/day.

Q. 30

Demonstrate the presence of phosphorus in the normal urine supplied.

Steps	Observation Points	Marks
1.	Take 10–12 mL of urine in a boiling tube add 3 mL of strong ammonia solution and boil till white precipitates of calcium and magnesium are formed.	1
2.	Filter through a filter paper placed in a funnel placed over a test tube. Wash the precipitate collected in the filter paper by just pouring a few mL of water through the filter paper.	1
	Take the funnel with the filter paper in situ, and place it over another test tube. Add 3 mL hot acetic acid through the filter paper placed over the test tube. Hot acetic acid with dissolved precipitate is collected in the test tube underneath. Divide it into 2 parts.	
3.	**To detect phosphorus:** Add a drop of concentrated HNO_3 and a few drops of ammonium molybdate solution. Boil → Fine lemon yellow (canary yellow) precipitate.	1
	Question	
1.	Mention the normal excretion rate and the source of it.	1 + 1

Ans. 30

1. *Urinary excretion rate of phosphorus:* 1g per day.
   Source: Phosphoproteins, nucleoproteins and phospholipids.

Q. 31

Demonstrate the presence of urobilinogen in the normal urine supplied.

Steps	Observation Points	Marks
1	Add 1 mL of Ehrlich's reagent (2% Para-dimethylaminobenzaldehyde in 20% HCl) to 10 mL of freshly voided urine.	1
2	Shake well and keep it in the rack for 5 minutes for the color development.	1
	Question	
1.	Mention the source of it.	1

Ans. 31

1. Heme.

Q. 32

Demonstrate the presence of indican in the normal urine supplied.

Steps	Observation Points	Marks
1.	To 5 mL of urine add 2 mL barium chloride and 2 mL hydrochloric acid. Mix well and filter.	1
2.	Divide the filtrate into two tubes. Boil the contents in one tube. Carefully look for the turbidity developing in the tubes.	1
	Questions	
1.	Mention the source of it.	1
2.	Name the condition in which it is excessively excreted in urine.	1

Ans. 32

1. Tryptophan.
2. Intestinal obstruction.

Q. 33

A man aged 60 years came with complaints of polyuria, polydypsia, polyphagia and loss of weight. Urine specimen of the patient is available. Do a simple test which will give clue about the disease. Write down specific tests to confirm the diagnosis.

Steps	Points to be observed	Marks
1.	Benedict's test.	1
2.	To 5 mL of Benedict's reagent taken in a test tube, add 8 drops of urine. Shake well and boil for 2 minutes or keep it in a water bath for 5 minutes → colored precipitate (green/yellow/orange/brick red).	1

Question	
1. What is the ADA approved diagnostic criteria of diabetes mellitus?	1

Ans. 33

Diagnostic criteria for diabetes mellitus: As per American Diabetes Association (ADA)*2015 guidelines Any one of the following is diagnostic of diabetes mellitus:
1. In a patient with classic symptoms of hyperglycemia(polyuria, polydypsia and polyphagia), a random (regardless of the time of the preceding meal) plasma glucose ≥ 200 mg% (11.1 mmol / L).
2. Fasting (no caloric intake for at least 8 hours) plasma glucose (FPG) ≥ 126 mg% (7mmol / L).**
3. 2-hour post load plasma glucose concentration ≥ 200 mg% (11.1 mmol/L) during an OGTT. **
4. HbA_{1c} > 6.5%.**, ***

*American Diabetes Association. Classification and diagnosis of diabetes.Sec.2. In Standards of Medical Care in Diabetes–2015. Diabetes Care 2015; 38 (Suppl.1): S8–S16.
** In the absence of unequivocal hyperglycemia, results should be confirmed by repeat testing.
*** HbA_{1c} testing should be performed with a method standardized to DCCT (Diabetes Control and Complications Trial) reference assay.

Q. 34

Raman aged 58 years, a farmer having the history of diabetes mellitus for 10 years taking irregular treatment came with vomiting, tiredness, disorientation and convulsions. What is your probable diagnosis? Do one simple test with the urine of the patient to establish your provisional diagnosis.

Ans. 34

Steps	Observation Points	Marks
1.	**Probable diagnosis:** Suffering from a complication of diabetes mellitus called diabetic ketoacidosis.	1
2.	Rothera's test to detect ketone bodies in urine.	1
3.	Saturate 5 mL of urine with ammonium sulfate crystals and add 2 drops of freshly prepared 2% sodium nitroprusside or a little of sodium nitroprusside powder.	
4.	Shake well. Add 1 mL of liquor ammonia through the sides of the test tube → Violet ring.	1

Q. 35

A boy aged 5 years was brought to the outpatient department with complaints of puffiness of face, edema of legs, reduced urine output. What can be the diagnosis? Urine of that child is supplied. Do the most appropriate test to support your diagnosis? What other blood test would be helpful to confirm the diagnosis.

Ans. 35

Steps	Observation Points	Marks
1.	**Probable diagnosis:** Nephrotic syndrome.	1
2.	**Heat and acetic acid test:** Fill three-fourth of the test tube with urine. Heat the upper one-third of the urine column by a small flame, so that lower two-thirds will serve as control.	1
3.	Add 2–3 drops of 1% acetic acid → White turbidity or coagulum.	1

More about nephrotic syndrome: Nephrotic syndrome is characterized by heavy proteinuria -Total protein > 3 g/24 h or albumin > 1.5 g/24 h, hypoalbuminemia, hypercholesterolemia and massive edema. Edema is the result of decreased oncotic pressure due to loss of protein. The transudation of salt and water into the interstitial spaces causes a decrease in plasma volume that in turn causes the kidneys to retain sodium. Abnormalities in lipid metabolism occur in the form of ↑ total cholesterol, ↑ VLDL, ↑ HDL. In spite of ↑ HDL cholesterol (HDLc), cholesterol remains high due to reduced activity of lecithin cholesterol acyltransferase activity (LCAT) thereby reducing reverse cholesterol transport leading to hypercholesterolemia.

Nephrotic syndrome results from a variety of causes—minimal change glomerulonephritis, membranous glomerulonephritis, drugs, infection, systemic lupus erythematosus, diabetic nephropathy.

Q. 36

A boy aged 10 years admitted with complaints of sudden onset of passing red colored urine, hypertension and pedal edema. The boy had a history of pyoderma on the skin about 2 months back. The urine collected from the boy is supplied. What is your provisional diagnosis? Name relevant tests with urine supplied and do one test.

Steps	Observation Points	Marks
1.	**Provisional diagnosis:** Acute glomerulonephritis (AGN).	1
2.	Name the urine test useful for diagnosis (heat and acetic acid to detect proteins and benzidine test to detect blood).	1
3.	**Heat and acetic acid test:** Fill three-forth of the test tube with urine. Heat the upper one-third of the urine column by a small flame, so that lower two-thirds will serve as control.	1
4.	Add 2–3 drops of 1% acetic acid → White turbidity or coagulum.	1
	Question	
1.	Give relevant investigations to be done in the blood in order to reach definitive diagnosis.	1

Ans. 36

More about AGN: Acute glomerulonephritis (GN) is characterized by the rapid onset of hematuria, proteinuria, reduced GFR and sodium and water retention leading to edema and hypertension. Many of the cases of acute GN are related to infection of skin or pharynx with group A β hemolytic streptococcal infection. This causes immune mediated injury to glomerular capillaries leading to GN. Immune mediated damage to glomerular barrier causes leakage of proteins into the urine and due to reduced GFR, nitrogenous waste products like urea and creatinine are retained in the blood causing uremia or azotemia. Investigations to be done in blood are estimation of urea and creatinine—these will be raised in glomerulonephritis.

Q. 37

A girl aged 15 studying in tenth standard came with complaints of headache, decreased appetite, yellowish discoloration of sclera and fever. On examination, the girl is febrile and jaundiced. Liver is slightly enlarged. Urine specimen of the patient is supplied. Give provisional diagnosis and do one relevant test in urine to establish jaundice.

Steps	Observation Points	Marks
1.	**Provisional diagnosis:** Viral hepatitis.	1
2.	**Relevant test in urine: Modified Fouchet's Test:** To 10 mL urine add 1 mL $MgSO_4$ and boil. While boiling add 10% $BaCl_2$ drop by drop till maximum precipitate is got. Filter and discard the filtrate.	1
3.	Take the filter paper from funnel and dry the precipitate on it by another piece of filter paper. After drying, add 2 drops of Fouchet's reagent to the precipitate on the filter paper.	1
4.	A bluish green color in the presence of bile pigments.	1
	Question	
1.	What blood tests you like to do to confirm the diagnosis?	1

Ans. 37

Blood tests:
  i. Serum total bilirubin (TB), conjugated bilirubin (CB), unconjugated bilirubin (UCB) are elevated.
  ii. Serum alanine transaminase (ALT) and serum aspartate transaminase (AST).
  iii. Serum alkaline phosphatase (ALP).

N.B.: In hepatic jaundice TB ↑, CB, UCB ↑.
Serum ALT and AST shows marked elevations.
Serum ALP shows slight to moderate elevations.

Q. 38

A 50-year-old man hailing from Attapadi of Palghat district came with complaints of joint pains and severe cramps in the legs, jaundice and hematuria. What is the provisional diagnosis? Do one urine test to diagnose this as a case of hemolytic jaundice?

Steps	Observation Points	Marks
1.	**Provisional diagnosis:** Hemolytic jaundice.	1
2.	**Relevant test in urine: Ehrlich's test.**	1
3.	Add 1 mL of Ehrlich's reagent (2% Para-dimethylaminobenzaldehyde in 20% HCl) to 10 mL of freshly voided urine → Red color.	1
	Question	
1.	What blood tests you like to do to confirm the diagnosis?	1

Ans. 38

More about this case: The history is strongly suggestive of hemolytic jaundice. Leg cramps, hematuria, joint pains along with jaundice are suggestive of sickling crisis related to sickle cell anemia.

Urine: Ehrlich's test for urobilinogen would be strongly positive and suggestive of hemolytic jaundice (increased rate of RBC break down → ↑ heme release and catabolism → ↑ urobilinogen in urine). Benzidine test for blood will also be positive. Hay's test for bile salts and Modified Fouchet's test for bilirubin will be negative since there is no obstruction to biliary flow in this case. In order to prove the presence of excess urobilinogen in urine due to hemolytic disease do Ehrlich's test. Do estimation of serum total bilirubin (TB), conjugated bilirubin (CB) and unconjugated bilirubin (UCB). The TB and UCB will be highly raised and CB will be marginally raised. Serum hepatic enzymes (ALT, AST and ALP) assay will show normal pattern since liver is not affected. Sickle cell disease can be confirmed by Hb electrophoresis to demonstrate Hb S band.

Q. 39

A 52-year-old fat lady came with abdominal pain, jaundice and itching.
Blood report is given below.
TB – 8 mg%; CB – 6.5 mg%; UCB – 1.5 mg%
Serum AST (SGOT) – 15 IU/L (Reference interval 8–20 IU/L)
Serum ALT (SGPT) – 30 IU/L (Reference interval 7–35 IU/L)
Serum ALP – 180 IU/L (Reference interval 25–75 IU/L)

The urine of the patient is supplied. Do two relevant tests to supplement the blood tests to arrive at a diagnosis. What is the cause of itching? Detect that substance in urine

Steps	Observation Points	Marks
1.	**Cause of itching:** Bile salt deposition in the skin.	1
2.	**Relevant test in urine:** Hay's test to detect bile salts in urine.	1
3.	**Hay's test:** Take 5 mL of urine in a test tube and sprinkle sulfur powder on the surface of urine → Sulfur powder sink to the bottom.	1
	Question	
1.	What blood tests you like to do to confirm the diagnosis?	1

Ans. 39

Blood tests:
 i. Serum Total bilirubin (TB), conjugated bilirubin (CB), unconjugated bilirubin (UCB).
 ii. Serum alanine transaminase (ALT) and serum aspartate transaminase (AST).
 iii. Serum alkaline phosphatase (ALP).

N.B.: In obstructive jaundice TB ↑, CB ↑, UCB–Normal or slightly elevated.
Serum ALT and AST normal or slightly elevated.
Serum ALP shows marked elevation.

Index

Page numbers followed by *f* refer to figure and *t* refer to table.

A

Acetic acid 44
Acetic anhydride sulfuric acid test 40
Acid base disorders 141
Acid of uric acid estimation 111*f*
Acid reagent 97
Acidosis, type of 143
Acrolein test 38
Acrolein, formation of 38*f*
ADH, secretion of 168
Alanine aminotransferase 118
Albumin 51, 52, 103, 104
    globulin ratio 105
    reactions of 30*t*, 32
    saturation tests of
        full 21*f*
        half 21*f*
Albustix 63
Alcohol-induced hypoglycemia 154*f*
Aldehyde test 28, 28*f*, 31
Alkali destruction test 15
Alkaline hypobromite 44
    test 43, 231
    and reaction 43*f*
Alkaline phosphatase 124, 125, 146
    isoenzymes of 126
Alkaline tartrate solution 91
Alkaloids 23
    type of 144
Alkaptonuria 157
Alpha-naphthol reaction 4
Amino acid 47
    color reactions of 25
    cysteine 29*f*
    groups of 27
    tryptophan 28*f*
Ammonia, test for 56, 59
Ammoniacal silver nitrate 49
Ammonium hydroxide 49
Ammonium molybdate solution 66
Ammonium sulfate
    crystals 21
    solution, saturated 20
Anticoagulated blood 72
Ascorbic acid 166

B

Barfoed's reagent 18
Barfoed's test 7, 7*f*, 12
Barium chloride solution 66
Beer-Lambert law 83
Benedict's reagent 6
Benedict's test 5, 6*f*, 7, 12, 16, 61
Benedict's uric acid
    reagent 49
    test 48, 48*f*, 232
Benign proteinuria 63
Benzidine test 63, 64*f*
Benzoic acid solution 97
Bile pigment 64
Bile salt 64
Bilirubin 114
    estimation 116*f*
    fate of 115*f*
Biuret method 104
Biuret reaction 31
Biuret reagent 106
Biuret test 25, 25*f*
Blood 63, 145, 149, 151
    specimen, collection of 79
    urea nitrogen 95
    urea, causes of high levels of 95

C

Calcium 128
    estimation 129*f*
    test for 57*f*, 59
Carbohydrate 3, 17
    classification 3
    disaccharides 3
    monosaccharides 3
    oligosaccharides 3
    polysaccharides 4
    reactions of 3
    substances 50, 62
Carbon dioxide 42
Carbonyl group, reactions to 4
Cardiac marker enzyme 171*f*
    activity 171*t*
Cardiac troponins 172
Cell division 128

Chloride, test for 56, 59
Cholesterol 39, 107, 109*t*
    color reactions of 40
    ester hydrolase 108
    estimation 107
    oxidase 108
    peroxidase 108
    structure of 107*f*
Clinistix 62, 62*f*
Cloudiness, causes of 54
Copper sulfate 6
Creatine 98
    kinase 172
    phosphate 98*f*
Creatinine 47, 51, 58, 98
    clearance 100
    reactions of 45
    test for 59
Cystatin C 170
Cystine 29*f*

D

Deoxyhemoglobin 69
Diabetes mellitus 53, 60, 85, 90*f*, 109, 149, 164
Diabetic ketoacidosis 145, 168
Diacetyl monoxime 94, 96
Dietary iodine deficiency 166
Dietary nucleic acids 110
Disaccharides, reactions of 11, 12
Disodium phenyl phosphate 126

E

Ehrlich's reagent 66
Ehrlich's test 60, 65, 65*f*, 148
Enzymatic methods 93
Erythropoietic porphyria 162
Ethereal sulfates, test for 60

F

False albuminuria 62
Fanconi's syndrome 112
Fats and fatty acids, reactions of 37
Fearon reaction 94
Ferric chloride 96
Filtrate, tests with 74
Fouchet's reagent 66
Fouchet's test 64, 65*f*
Foulger's test 9, 9*f*, 14
Fructose 50-52

G

Galactosemia 157
Gelatin, reactions of 34
Globin hemochromogen 71, 73, 208
Glomerulonephritis 169
Glucose 51, 52, 61, 85
    concentration 85
    oxidase method 86, 86*f*, 87
    solution 18
    tolerance test 88
Glyoxylic acid 28
Grease spot test 38

H

Halogenation test 39*t*, 231
Hay's test 64, 64*f*
Hb derivatives by spectroscopy 70*f*
Heart disease 122
Heat coagulation test 23, 62, 62*f*
Heller's test 25*f*
Hemin crystals 71, 72*f*
    preparation of 71
Hemoglobin derivatives 67
Hemoglobin pigment 67
Hepatic disorders, ALT and AST in 122
Homopolysaccharide 224
Hopkins-Cole reaction 28
Human chorionic gonodotropin 74
Hypercalcemia 131
Hyperphosphatemia 136
Hyperuricemia 112
    causes of 112
Hyponatremia, symptoms of 167
Hypophosphatemia 135
    causes 135
Hypouricemia 112
    causes 112

I

Indoxyl sulfuric acid 58*f*
Iodine test 16, 16*f*
Iron deficiency anemia 163
Isotonic sodium sulfate-copper sulfate solution 91

J

Jaffe's reaction 99, 99*f*
Jaffe's test 46, 46*f*, 232
Jaundice 117, 145
    hepatic 117
    obstructive 109
    posthepatic 117
    prehepatic 117
    types of 117

K

Ketone bodies 63
Kidney 143-145
    disease 168
Kupffer cells in liver 59

L

Lactate dehydrogenase 172
Lactosazone 13f
Lactose 11, 50-52
    solution 18
L-amino acids 103
Lead blackening test 29
Libermann-Burchard reaction 40, 41f
Lipids
    classification of 37t
    reactions of 37
Lohmann's reaction 98, 98f
Lung 143, 145

M

Maltosazone 13f
Maltose 50-52
    solution 18
Metabolic acidosis 145
Metabolic alkalosis, causes of 144
Metabolism, inborn errors of 155
Metaphosphoric acid 23
Methemoglobin 70
    solution 73
Methionine 29f
Methyl guanidinoacetic acid 98
Milk, reactions of 74
Millon's test 27f, 28, 31
Mineral acids 24
Molisch test 4, 5f, 12, 15
Molisch's reagent 18
Molybdic acid reagent 136
Monosaccharides
    classification of 4t
    for osazone formation 9t
    reactions of 4
Murexide test 48, 49f
Myocardial infarction 170
Myoglobin 171

N

Nephrotic syndrome 109, 169
Neumann's test 229
Ninhydrin test 26
Nitric acid 66
Nitroprusside test 45
Noncarbohydrate substances 62
Nonprotein nitrogen 45, 47

O

Oliguria 53
    causes of 54
Oral glucose tolerance test 89, 90f, 91f
    types of 89
Organic constituents, tests for 61t
Organic solvents 23, 24
Organic sulfates 58
Orthophosphoric acid 96
Orthostatic albuminuria 63
Osazone test 9, 10f, 13
Oxyhemoglobin 69, 72

P

Pauly's test 30, 31
    cherry red 29f
    orange red 29f
Peroxisomal phytanic acid oxidase 166
Phenol red indicator 44
Phenyl ketonuria 53
Phenylhydrazine mixture 18
Phenylketonuria 157
Phosphate 133
    test for 59
Phosphomolybdic acid reagent 92
Phosphoric acid 124
Phosphotungstic acid 111, 112
Photoelectric colorimetry 80
Picric acid
    reaction 46, 232
    solution 101
Plasma glucose 89
Plasma pools
    of calcium 129f
    of phosphate 133f
Polysaccharides, reactions of 15
Polyuria 54
    causes of 54
Porphyrias 161
Potassium ferricyanide 127
Potassium oxalate 66
Protein 62, 103
    color reactions of 25
    estimation 103
    free filtrate 111
    functions of 103
    housekeeping 19
    reactions of 19
    standard stock 106

R

Reagent preparation 18, 44
Respiratory acidosis 143
Rothera's test 63, 63*f*

S

Sakaguchi's test 29, 31
Salkowski's reaction 40, 40*f*
Schiff's test 48, 48*f*
Seliwanoff's reagent 18
Seliwanoff's test 8, 8*f*, 13
Serum albumin 105, 115
Serum bilirubin values 146
Serum creatinine 170
Serum enzymes 146
Serum globulins 105
Silver nitrate solution 66
Sodium
    bicarbonate 126
    bisulfite solution 136
    carbonate 6, 112, 126
    chloride 106
    dithionite 69
    hydroxide concentrated 44
    hydroxide solution 101
    hypobromite 44
    potassium tartrate 106
    sulfite solution 136
    tungstate 91, 101, 112
Spectroscopes, types of 67
Starch hydrolysis test 16, 16*t*
Starch, reactions of 15
Sucrose 11, 50-52, 224
    solution 18
    test 14, 14*f*, 43, 43*f*
Sugar 12
Sulfonylureas, insulin 88
Sulfur test 29, 29*f*, 31
Sulfuric acid 96, 136
Sulfate, test for 56, 59

T

Thiosemicarbazide 96
Trichloroacetic acid 136
Tyrosine and tryptophan 27*f*

U

Urates in urine 54
Urea 51, 52, 57, 93
    formation of 42*f*, 93*f*
    reactions of 42
    solution, properties of 42
    test for 59
    urease action on 43*f*
Urease enzyme 44
Urease solution 66
Uremia, causes of 96*f*
Uric acid 47, 47*f*, 51, 52, 57, 110
    estimation 110
    formation of 110*f*
    molecules 47
    reactions of 47
    stock standard 113
    test for 59
Urine 146, 149, 151
    abnormal constituents of 60
    analysis 53
    constituents of 53
    creatinine determination 101
    excretion in 60*f*
    inorganic constituents of 55
    normal 55*t*
    organic constituents of 57, 59
    sample collection 80
    tests for inorganic constituents in 59*t*
Urobilinogen 59, 65
    formation 60
    test for 60

V

van den Bergh reaction 117
    direct van 115
    indirect 115
Visual colorimeters 80
Vitamin
    and minerals 163
    C 166
    D 107
        deficiency 165
    K deficiency 165
von Gierke's disease 156*f*

W

Water and electrolytes 167
Whipple's triad 153
White blood cells 54
Wilson disease 164

X

Xanthoproteic reaction 27, 31, 228
Xanthoproteic test 27*f*

Z

Zak's method of cholesterol estimation 108*f*